The Promise of Access

T0176009

The Promise of Access

Technology, Inequality, and the Political Economy of Hope

Daniel Greene

The MIT Press
Cambridge, Massachusetts
London, England

An early version of "Discovering the Divide: Technology and Poverty in the New Economy" first appeared in the *International Journal of Communication* 10 (2016).

This book was set in Stone Serif and Stone Sans by Westchester Publishing Services. Printed and bound in the United States of America.

Library of Congress Cataloging-in-Publication Data

Names: Greene, Daniel, 1973- author.
Title: The promise of access : technology, inequality, and the political
    economy of hope / Daniel Greene.
Description: Cambridge, Massachusetts : The MIT Press, [2021] |
    Includes bibliographical references and index.
Identifiers: LCCN 2020025454 | ISBN 9780262542333 (paperback)
Subjects: LCSH: Digital divide--Washington (D.C.) |
    Computer literacy--Social aspects--Washington (D.C.) |
    Knowledge economy--Washington (D.C.) | Poverty--Washington (D.C.) |
    Technology and state--Washington (D.C.)
Classification: LCC HM851 .G7446 2021 | DDC 303.48/3309753--dc23
LC record available at https://lccn.loc.gov/2020025454

10 9 8 7 6 5 4 3 2

Contents

Acknowledgments   vii

Introduction: "The Internet: Your Future Depends on It"   1

1   Discovering the Divide: Technology and Poverty
in the New Economy   29

2   The Pivot and the Trouble with "Tech"   59

3   "More Than Just a Building to Sit In for the Day"   81

4   Flexible Classrooms   111

5   Bootstrapping   141

Conclusion: Reproducing Hope   171

Notes   193
References   207
Index   233

Acknowledgments

I've got a really big team. And I'm not much without them.

I must first thank my informants and everyone else who allowed me into their offices, labs, and lives in support of this project. Without their generosity, none of this would be possible. I owe special thanks to those generous people who not only welcomed me in, but provided feedback on this manuscript and its argument over the years: Shawn, Ebony, Beth, Catherine, Irene, Daniella, Grant, and Eugene. Donna Haraway (1997) tells us that ethnography "is not so much a specific procedure in anthropology as it is a method of being at risk in the face of the practices and discourses into which one inquires" (190). It is from my informants that I learned I was not studying the digital divide per se but the institutions managing it, ones in which I myself was deeply enmeshed. Tracking the hope that animates and renovates these institutions required deep questioning of my own professional and political identity. For that I will be eternally grateful.

This project emerged from my dissertation in American studies at the University of Maryland. The members of my committee each shaped this project and my thinking in their own way and so I thank Ira Chinoy, Sheri Parks, Jan Padios, Katie Shilton, and Jason Farman. Other mentors, inside and outside my American studies doctoral program, also provided invaluable support and perspective: Christina Hanhardt, Lisa Nakamura, Matt Kirschenbaum, Jessie Daniels, Alice Marwick, and André Brock. Without Patrick Grzanka's friendship and encouragement, I doubt I'd even have started graduate school, much less finished it.

Without my brothers, sisters, and comrades in the University of Maryland's graduate workers' union, or our allies in the staff's American Federation of State, County, and Municipal Employees (AFSCME) local or the undergraduates' Student Labor Action Project, I wouldn't have been reminded of my priorities throughout the research for this project, nor would I have been able to recharge and apply myself to the things that I really want to make a career of: making our institutions work for everyone in them and making sure our students' learning conditions don't suffer for our working conditions.

The Social Media Collective at Microsoft Research New England provided the best possible space in which this project and its author could mature. Many thanks to the generosity of Nancy Baym, Tarleton Gillespie, Mary Gray, Kate Crawford, danah boyd, and the other friends and mentors I met in that lab during my postdoc. Many thanks too to my brave coworkers in big tech, who, in the spring of 2018, rebelled against their employers, joined with the fighters at Color of Change and Mijente, and organized in defense of the people targeted by the surveillance and classification systems that Google, Microsoft, Amazon, and others built to help governments bomb or cage people. I did not believe such a thing was possible and I believe that you will win.

A wonderful network of wonderful people supported me and my work at important times, providing guidance, advice, and camaraderie as this book progressed. They include Daniel Joseph, Gavin Mueller, Deen Freelon, Ifeoma Ajunwa, Luke Stark, Anna Lauren Hoffman, Mike Casiano, Katy Pearce, Mary Flanagan, Meg Finn, Jessa Lingel, Nick Seaver, Tara McPherson, Miriam Posner, Simone Browne, Karen Gregory, Brooke Duffy, Douglas Williams, Joanna Pinto-Coehllo, Leslie Kay Jones, Shannon Mattern, Lilly Irani, and Steve Vallas. I would also thank the Relaxed Marxist Discussion Group for good conversation. Gita Manaktala is the best editor I could have asked for as a first-time author. Laura Portwood-Stacer and Sarah Hamid provided invaluable services in reviewing and preparing the manuscript. Special thanks are owed to those colleagues who provided feedback on the manuscript as it neared completion: Jen Jack Gieseking, Victor Ray, and Nick Seaver.

I write these acknowledgments as a pandemic sweeps the world. This is ultimately a book about care, about making a living and making lives. The biggest thanks are due to my daughter, Eliza: care is a promise, for a world we'll build together. And to my partner, Annie, who continues to care for families and children no matter what and who has taught me more about this fight than anyone else.

Introduction: "The Internet: Your Future Depends on It"

In 2013, a series of posters started appearing across Washington, DC. Each one declared "The Internet: Your Future Depends on It," next to a photo of a Black Washingtonian. "Sean earned an advanced certification in six months. Now he upgrades computer systems for the US Small Business Administration. He uses technology to help people start businesses. So can you." They looked ahead to their new future, smiling. "Fabiane learned Microsoft Office in eight weeks and used her new skills to write, design, and publish her first book. She's using technology to pursue her dreams. So can you."

They told their stories about using digital training resources provided by Connect.DC—the DC municipal government's technological assistance program—to get to that future. These skills and tools led to better jobs, ones in which you don't work with your hands. "Selina's computer certifications helped her get a job with the federal government—after spending 20 years as a beautician. She used technology to change careers." The jobs were often in the federal government, which was not surprising. DC remains a company town in many ways. Federal employment is the foundation of the regional labor market (Office of Revenue Analysis 2017), and of the region's Black middle class (Lacy 2007). But any regional labor market is vulnerable when it depends on a lone employer, especially one with a budget that's constantly under attack. Unsurprisingly, the DC municipal government has long sought to lessen the city's reliance on federal employment.

These posters built a link between individual skills training and the city's economic growth. The rewards were bigger than any one person's career. It was not just a matter of bringing single individuals across a *digital divide*, a gap between those who had internet access and the skills to use it and those who didn't. The dream here was bigger: by changing their tools and skills, people could change the community in which they lived. "Marcus earned three computer certifications in less than a year. Now he works as a computer technician with the DC government. He uses technology to improve his city."

Curiously, the internet itself wasn't mentioned in these testimonies. No one was designing websites or setting up e-commerce portals. As a network of networks, the internet, in these posters, was so rhetorically expansive as to stand in for any digital technology (e.g., Microsoft Office) and any credential that evidences any digital skill (e.g., advanced certifications). The internet in "The Internet: Your Future Depends on It" is not a specific tool but a symbol of economic progress—the promised land you'll reach with the right equipment and the right training.

The portraits were empowering. The people in it used technology; it didn't use them. They were inviting. They shared successes and asked you to join them: "So can you." Readers were told to "text **LEARN** to **83224** for information about **free** technology and job training." The tools were there—you could even get them for free!—but time was of the essence. There is a familiar, attractive story here: learn to code, get online, secure your future. The question is: Why do we understand economic success and failure in these terms?

Across the city, people were grappling with this promise in different, more complicated ways. During the research for this book, I was lucky to meet Ebony and her boyfriend Shawn, Black Washingtonians in their early twenties.[1] Generous, funny, and inseparable, they spent most of their days in the computer lab of the public library, first at the Northwest One branch and then, after an ugly confrontation with library police for "sleeping," at the larger Martin Luther King Jr. Memorial Library—Central Library (MLK) a few blocks south. Ebony eventually went back to school for her GED. After classes ended for the day, she would return to the lab, hug Shawn, and catch up with the rest of their friends. Once the library closed, they often walked over to a nearby Subway restaurant—not to order a sandwich, but to sit outside and use the free internet for a few hours, browsing Facebook, watching cartoons, and reading before they returned to separate shelters for the night. Their friend Josie had a similar spot, an outlet above a Metro escalator where she could get Wi-Fi from the nearby basketball stadium. She could listen to music as the crowd of White suburbanites in town for the game walked around her.

For Ebony, Shawn, and Josie, the internet was certainly an important part of their everyday lives—but it wasn't the promised land of the future. It was the terrain of the present. Shawn said that when he first lost housing after he and his father were evicted, "the library became my best friend." It's where he made friends, where he got online to chat or play games or follow the news coverage of the Occupy DC encampment he worked on. "I was always a computer man," he told me.

There is a whole story in "I was always a computer man," a story about Shawn's hobbies, his friends, and the welfare state institutions through which he traveled. He was not left behind by technological or economic progress, as the posters might suggest. Rather, he used the PC and the institutions around the PC to build a rich and varied life and to claim space in a gentrified city increasingly hostile to the working and workless poor. Shawn was a "computer man," an activist, a homeless man, a boyfriend, a library patron, an artist, and more. "I was always a computer man" is both a humble biographical fact and a political statement about the security he felt behind a screen in the library—where Shawn was welcomed without having to purchase anything, where he could find peace without having to prove his economic value. The PC was not a skills engine for him; it was one piece of a place-making project. While he spent most of his time at MLK, Shawn still claimed the Shaw branch as his own: "That's my library." The decision to move from one library to another was based on whether he felt those sorts of claims to public space were being respected by the institution.

"I was always a computer man" is a longstanding identity that supports a claim to public spaces and public resources that belong, or should belong, to everyone. Claiming these spaces and resources can be a great deal of work: Josie worked hard to find a Wi-Fi signal at night, and Ebony and Shawn made tough decisions about which libraries were safest for them. But they were owed these spaces in the present. Where they did not find welcome, they still made space for themselves.

"The Internet: Your Future Depends on It" is a different story. It has a different temporal orientation, rearranging present institutions based on a future threat to individuals. And it has a different calculus for social worth, arguing that your social value will diminish with your economic value. It is also a familiar story. Both liberal and conservative politicians have repeated it for decades now, updating it for the technology of the day in an attempt to persuade the public that their individual and collective economic futures depend on their access to the right skills and tools:

- On the campaign trail in 1996, Bill Clinton asked a crowd in Tennessee to "keep faith with our future by passing on to our children an Information Superhighway that will help them to live out their dreams" (Clinton and Gore 1996b).

- In 2004, George W. Bush campaigned against Clinton-era taxes on consumer internet that he said restricted broadband deployment: "It's the flow of information and the flow of knowledge which will help transform America and keep us on the leading edge of change. And we've got to make sure that flow is strong and modern and vibrant" (Bush 2004).

- In 2013, President Obama celebrated Computer Science Education Week with a video imploring young people to learn to code, not just for the sake of their future but "for our country's future," saying: "Don't just buy a new video game. Make one. Don't just download the latest app. Help design it. Don't just play on your phone. Program it" (Obama 2013).

- Commenting on the Education Department's STEM grants, White House official and first daughter Ivanka Trump said in 2017, "Given the growing role of technology in American industry, it is vital our students become fluent in coding and computer science, with early exposure to both" (Romm 2017).

- On a campaign stop at a Black-owned barbershop in Des Moines, Iowa, in 2019, presidential hopeful Kamala Harris suggested an undergraduate pursuing a career in law should learn to code. She then gave the same advice to a young woman majoring in political science (Enjeti 2019).

"The Internet: Your Future Depends on It" is then just one local example of a story that explains individual, regional, and national economic success as a product of information technology and the skills to use it. The DC city government's message is part of this political common sense. Political common sense mixes factual claims with ideological ones. It is clear, for example, that finding a job without an internet connection and a PC is difficult; just try filling out an application to work for CVS, let alone using USAJobs, on your phone (Smith 2015). It is less clear that there are plenty of good tech jobs out there, just not enough coders to fill them. In the course of my research in DC Public Library (DCPL) branches, I met not just patrons like Ebony and Shawn but librarians like Becca or Grant who worked with them. These helping professionals readily acknowledged that economic reality is more complicated than political slogans might admit; getting online and learning to code won't change the rest of the labor market by itself. But they and their organization still embraced these claims about the power of internet access and digital skills training. They had to try to help. The stakes were high. The future depended on it.

This book investigates how the problem of poverty is transformed into a problem of technology and how the organizations addressing the problem are themselves transformed in the process. It asks how "The Internet: Your Future Depends on It" becomes common sense. And it asks what happens to people like Shawn, whose stories—"I've always been a computer man"—do not fit that common sense. Spending time with the people and places that act on poverty with technology, we can begin to understand the attraction of this simple, powerful story—even for those who know reality is more complicated. Indeed, as the skyrocketing inequality of the information economy has become harder and harder to ignore, "The Internet: Your Future Depends on It" has become more and more attractive.

I call this political common sense the *access doctrine*. The access doctrine decrees that the problem of poverty can be solved through the provision of new technologies and technical skills, giving those left out of the information economy the chance to catch up and compete. The access doctrine emerged from 1990s debates over the problem of persistent poverty in a globalizing, deindustrializing US economy whose most profitable frontier seemed to be based in using and producing information and communication technologies. Schools and libraries threatened by fiscal austerity or accusations of obsolescence embrace the access doctrine as their mission in order to restore their legitimacy, garner much-needed resources, and simplify the host of social problems with which they are confronted daily.

Common sense is never a totally coherent, factual account of the world. Nor is it a deception perpetrated by the ruling classes on the ruled (Gramsci 2000). Rather, common sense emerges organically from practical responses to real problems in the real world, crafted from the symbols and materials at hand. A world with clear and discomfiting divisions between ruler and ruled. A capitalist world, where the major institutions of social life necessarily reproduce these divisions, not in a mechanical fashion but because places like schools are built to last for generations and so must establish the normative outlines of the world they wish to make, and because they too must make sense of the mess of social life confronting them.

The access doctrine helps us make sense of the world—specifically, a city like so many others, where the technology sector expanded as part of a wave of post-2008 high-wage, White emigration, while Black unemployment remained persistently high, homelessness increased, and working-class wages barely budged. The access doctrine makes these problems appear natural and immutable, like an earthquake, and teaches individuals and organizations how to survive them. Our task then is to separate myth from fact, ground that precise mix of myth and fact in the historical process, and rearrange the incoherent story into a coherent explanation of how the world works, why it works that way, and why we believe otherwise.

Skills Gaps and Digital Divides

The access doctrine offers clear solutions. But it is far from clear that consumer internet access, digital skills training, or skills training generally produce widespread economic mobility. To investigate the specific political utility of the access doctrine, we must first clarify the general problem it is trying to solve and whether previous attempts to solve that problem have succeeded.

In the 1990s, the story went that access to certain tools—an internet connection and a PC—would enable access to high-wage jobs. Today, it appears that the tools were easier to get than the jobs. Ultimately, we must reckon with the fact that as home internet access has reached a saturation point—though of course gaps still remain[2]—the economic lives of working Americans have either not improved or have worsened. Hourly wages have stagnated since the late 1970s, while productivity has continued to improve (Bivens et al. 2014).[3] Indeed, from 2000 to 2013, the era wherein home internet access reached the saturation point, hourly wages for 30 percent of the workforce actually fell. The story is worse for workers with only a high school degree, but even college graduates have seen anemic wage growth. And while gender-based wage gaps have narrowed over the last forty years, race-based ones have been stubbornly persistent (Bivens et al. 2014; see also Akee, Jones, and Porter 2019; Autor and Dorn 2014; Duménil and Lévy 2015; Edelman 2013). This is hardly the picture we'd expect if internet access increased individual economic fitness.

But the access doctrine is expansive. As more consumers got online, the story changed: It was not just access to specific tools that were necessary for economic mobility, it was specific digital skills—the sort of things advertised in the Connect.DC posters. Shawn, this story would say, had access to the right tools, but not the right skills. This story is part of a wider genre that explains the relative poverty or economic inactivity of specific populations, regions, and countries through the skills individual workers possess. Chapter 1 will review how this problem became a problem, but for now it suffices to say that for much of the twentieth century, especially after the economic dislocations of the 1970s, representatives of business, education, and government have warned of impending or actual *skills gaps* (i.e., where the education system does not provide graduates with the skills businesses need) and the more specific problem of *skills shortages* (i.e., where US businesses cannot find a specific kind of worker—today, generally engineers and information technology professionals) (Cappelli 2015). The access doctrine emerged from these earlier frameworks, adding to them the new technologies and skills associated with the internet.

Skill is notoriously difficult to define and measure, so educational credentials and professional certifications are generally used as a proxy in the measurement of skills gaps or shortages; within firms, self-reported data on skills needs, rather than independent assessment, is the norm. The urgent problem of skills gaps and shortages is given historical weight through the invocation, typically by economists (e.g., Autor, Katz, and Krueger 1998; Goldin and Katz 2008; Krueger 1993; Tinbergen 1974), of skill-biased technological change (SBTC). Advocates for SBTC hold that the prevalence of *technology*—itself typically unmeasured and underdefined—increases in prevalence and complexity over time, and thus the demands for and wage returns of workers skilled in its design and use also rise over time. The vagueness of *skill* and *technology* are empirically troublesome but politically useful: a nebulous threat remains always on the economic horizon, explaining that those struggling today do so because they have not sufficiently upgraded—and that they must do so before tomorrow arrives. So while I present some objections to policies based on SBTC here, I do so knowing that, politically, it's ultimately unfalsifiable. Even if today's training regime does not work, new technologies will always arrive and, for those interested in shifting the risk of economic transition to individual workers, they will always demand new skills. The critiques presented here simply lay the groundwork for the later investigation of public service organizations animated by their faith in SBTC.

A theory of skill-biased technological change demands a policy of technological training. Unfortunately, the returns from dedicated, large-scale skills-training programs have been mixed at best. The largest such experiment was President Reagan's replacement of the Comprehensive Employment and Training Act (CETA), which gave funds to local governments to provide full-time jobs to the poor and long-term unemployed, with the Job Training Partnership Act (JTPA)—later transformed into the Workforce Investment Act (WIA). JTPA replaced CETA's guaranteed jobs with training courses (Smothers 1981) and enrolled close to a million people during the Reagan and Bush administrations (Lafer 2002). It offered fast-track trade training in classrooms, supplemented by on-the-job training and job search assistance. The smaller Trade Adjustment Assistance program provided similar services for laid-off manufacturing workers.

In reviewing their results, Lafer (2002) found that these programs rarely covered more than 5 percent of the affected populations. The Labor Department's own review of JTPA (Bloom et al. 1997), based on a two-and-half-year controlled experiment, found that only a quarter of JTPA's participant categories saw relative income gains. The study found no effect on participants' welfare benefits because the income gains were not enough to bring them above the federal poverty line. The impact of training on earnings remained small relative to the impact of gender (gender differences accounted for four times the earnings difference of training). Adult men and women participants saw small net gains to their earnings, while youth saw substantial losses, and the net social benefit (i.e., program cost relative to participant returns) largely zeroed out.

These interventions may have failed because they target the wrong skills. It is important to note that defining and measuring workers' skills, and understanding what employers mean by *skill*, is quite hard: it can mean tasks or abilities, it can be rewarded or not, and sometimes it appears to be just be a proxy for race or gender (Attewell 1990; Steinberg 1990). Nonetheless, labor sociologists who attached occupational skill measurements to returns from the Current Population Survey in 1979–2010 found that it was analytic, critical thinking skills that produce the greatest wage payoffs—not the technical skills that are supposedly so in demand (Liu and Grusky 2013). Similar research, reviewing returns from the General Social Survey in 1972–2002, the period in which the access doctrine was born, found that overqualification (excess educational credentials relative to job tasks) increased substantially in the exact period when these credentials were supposed to be more in demand. Job satisfaction unsurprisingly fell during this time (Vaisey 2006). So not only has upskilling to meet the demands of skill-biased technological changed probably failed—or at least not worked as we were told it would—it has led to a rise in overqualification and job dissatisfaction.

Despite this, the push for technical skills training continues. In 2015, President Obama announced the $100 million TechHire initiative to provide grants to local public-private partnerships that would provide technological skills training. In announcing it, he warned of a crisis of national economic fitness: "If we're not producing enough tech workers, over time that's going to threaten our leadership in global innovation, which is the bread and butter of the 21st century economy" (Kuhnhenn 2015). That crisis was disputed within his own administration. That same year, the Bureau of Labor Statistics (BLS 2015a) released a report showing that because capital moves faster than the labor it employs, any skills gap is necessarily local and temporary—and that local oversupply of professionals with a particular skill set is as much a possibility as undersupply.

Nationally, one would expect the wages of degree earners in science, technology, engineering, and math (STEM) to have boomed in times of high demand, like those Obama described. But median STEM wages have remained stable since 2000, and most STEM degree holders work in non-STEM fields—in part because firms are able to outsource high-skilled labor to low-wage workers through digital networks or import them through programs like the H-1B visa system (Charette 2013). The BLS (2015b) also found that the fields of projected high job growth in the next two decades are in low-wage service work like home health care and food preparation, not high-wage knowledge work. For most workers, the jobs of the future rely on catheters and cutting boards, work that requires no knowledge of Python or JavaScript and often no more than a high school education. Should this labor market mix remain stable, hypercompetition for a few high-wage jobs and overqualification for the remaining jobs will be the norm (Berghel 2014).

The geographic version of the skills gap argument holds that redesigning a particular region to court technical "creative class" emigres will create a rising tide that lifts all boats (Florida 2004). This, too, has been thoroughly contested. Critics identify the emigration of cosmopolitan, high-wage professionals into urban areas with rising inequality and the displacement of working-class residents of color (O'Callaghan 2010; Peck 2005, 2007). Economic incentives offered by states to the firms that might relocate, change the local skill mix, and move local economies up the value chain generally have the effect of increasing local inequality by allowing wealthy firms and individuals to keep revenue that would otherwise be taxed (Jansa 2020). And at a national level, no more prominent a figure than longtime chair of the Federal Reserve Alan Greenspan, who perhaps more so than anyone else built the US information economy, admitted in 2007 that the main effect of importing skilled professionals is to "suppress the skilled-wage level" (Bloomberg News 2007).

There appears, then, to be a deeper problem than what skills we pick to measure and what returns they may or may not deliver to individuals, regions, and countries. The problem may be framing economic mobility—of individuals, cities, or countries—as a product of *individual* skills. Cappelli's (2015) extensive research into the gap between the skills employers say they need and what they actually utilize shows this conceptual problem playing out on several levels. First, the problem differs depending on who you ask. Employers say they don't have enough skilled workers. Professional associations of skilled workers say that the problem is in fact oversupply. Recruiters, brokering between the other two parties, say the missing ingredient isn't certain skills but certain attitudes, such as conscientiousness. Second, employers change their skill requests based on fluctuations in the labor market, raising them when it's loose and, when it's tight, either lowering skill requirements or automating higher-skill tasks. Finally, employer complaints consistently focus on new labor market entrants—young, relatively lower-paid workers who make up a tiny minority of the total workforce. This seems to indicate an unwillingness to either train new workers or raise the wages of skilled veterans; unsurprisingly, lateral movement of mid-career workers between firms has increased since the 1970s (Cappelli 1999).

Regardless of the source of its conceptual confusion, the skills-gap narrative shifts the responsibility for training away from the business owners and toward young individuals at the beginning of their working lives, as well as toward the public institutions that train them: universities, community colleges, and local and state governments. Indeed, what little evidence we have suggests that employers today devote very little time to on-the-job training, relative to the postwar "golden era" of long-term, single-firm employment (Cappelli 2015). At the level of statecraft, placing a political priority on skills training is a positive, forward-looking way to justify cuts to unemployment insurance—because it's unnecessary if there really are enough jobs available for the skilled—and the general shift of training burdens from firms to schools, libraries, and individuals. Whether or not skills training works as a solution to individual or regional economic immobility, it does work in the sense that places like schools and libraries and the people who staff them are reorganized to attack these new priorities.

So the evidence for the existence of a skills gap is at best mixed, as is the evidence of positive labor market returns, for individuals or regions, from attempts to skill up local workers or import new ones. But still, the hope survives. These problems are complex, but the access doctrine continues to make sense as a simple solution. It's easy to see that in statements from President Obama, who insisted throughout his term that "in the new economy, computer science isn't optional—it's a basic skill" (Obama 2016). These sorts of warnings about the nationwide need for more computer scientists are common. They condense into a three-word incantation that insists on a surefire path to economic security and threatens obsolescence for the unskilled: "learn to code" (Miltner 2019; Williamson 2016).

"Learn to code" is a relatively recent creation of the access doctrine, decrying skills shortages in in software development. Decades prior to "learn to code," Vice President Al Gore (1994c) warned of a future gulf between "information haves and information have-nots." Both formulations understand poverty as a *digital divide*, itself a term that has waned in popularity even as the access doctrine has remained strong. Digital divide researchers have spent decades demonstrating that the problem is more complicated than a simple binary between haves and have-nots, coders and noncoders; getting people online or teaching them to code is not a shortcut to economic security (Gunkel 2003; Stevenson 2009). But we social scientists find ourselves confronting these simple binaries again and again. They persist, I argue, not because they accurately reflect the state of the world, but because they are politically useful.

Take Mia, for example, whose life conforms to no simple binary between information haves and have-nots. Friend to Shawn, Ebony, Josie, and nearly everyone else in MLK's computer lab, Mia occupied a regular seat in the back rows of the Digital Commons, where she used her glitter-coated laptop, a twenty-fifth-birthday present from her grandmother, to search for permanent housing for herself and her mother, find work, rest between appointments, work on her manga, and watch movies and TV. In terms of teaching people how to phrase a Google search or look up the location and hours for a particular government office, she was as important to the library as any librarian. She was a helper too, not just someone who was helped, and her talents disrupted any idea of a stark digital divide between those with skills and access and those without.

Currently homeless, Mia struggled to secure a regular internet connection but knew one could usually be found at the library and had plenty of back-up locations at the ready. She was working toward a certification in medical billing at the University of the District of Columbia, but multiple disabilities and the daily disruptions of homelessness interrupted both education and job search. Her phone and laptop were always close to hand, but both regularly had their service disrupted or their physical casings damaged (Gonzales 2016).

"I've always been a big library person," Mia told me, echoing Shawn's claim that he'd "always been a computer man." Digitally literate but still suffering from the high costs and daily stress of a gentrifying city, her life didn't fit the access doctrine or, as we will see, the vision of the library remade to support the access doctrine. While digital divide research has increasingly reckoned with the complexity of digital experiences like Mia's, this urgent belief in a gulf between the skilled and wired and the unskilled and unconnected persists in the media, in our politics, and in our institutions tasked with confronting poverty. The access doctrine could only understand Mia as someone to be helped. It could only offer tools and training in support. But she already had the tools and she was getting the training. What she needed at the time was a safe, permanent place for her and her mother to live.

The access doctrine has a political gravity that draws us all in, whether we embrace it or refute it. It has a quality of "magical thinking," as Eubanks (2011) writes in her critique of "information poverty" approaches to education and economic development. For Eubanks, access is not a question of whether a skill or technology is present or absent, but a question of citizenship and the capacities for engagement in public life that emerge from a life under surveillance. She is right, of course, but the magical thinking persists and is seemingly immune to attempts, both empirical and conceptual, to debunk it.

State and local governments continue to bet on this magical thinking, despite the heavy burden it places on institutions ill-equipped to solve it. High schools and colleges, for example, have a hard time adjusting multi-year curricula to keep up with ever-changing employer demands, preparing adolescents who do not yet know who they are for a labor market that does not yet exist. But, as we will see in chapters 3 and 4, libraries and schools eagerly embrace this challenge, even if it is beyond their abilities. They do so because public service organizations are themselves under threat in an environment of overwhelming economic uncertainty and inequality. By turning the problem of poverty into a problem of technology, they reframe their own problems into something more manageable for their frontline staff and more legible to the politicians, donors, and others who might offer support. These institutions teach us how to survive, but their own survival hinges on reproducing the common sense of the access doctrine. That inter- and intraorganizational dynamic is the focus of this book.

Features, Not Bugs

So it seems technology provision and skills training don't solve poverty or inequality—or at least the story is much more complicated than that. Why then do these solutions remain economic cure-alls, embraced across the political spectrum? Rather than trying to disprove this idea of a simple digital divide, this book offers an immanent critique of it, exploring first the historical origins of its premises and then how that logic is reproduced day to day, failing on its own terms, but surviving nonetheless. This simplicity is a feature, not a bug, offering tremendous utility to those who embrace it. The access doctrine pushes the institutions that manage the problem of poverty—here, schools and libraries—to *bootstrap*, remaking their identities and operations to support technology provision and skills training. Bootstrapping secures resources in periods of fiscal austerity and legitimacy in periods of political assault.

Like Eubanks, most of the teachers and librarians I speak with throughout this book know the real story is more complicated than any magical thinking might suggest. They know full well that the facts of poverty, racism, homelessness, and disability cannot be overcome with a laptop and a broadband connection. But this story is nonetheless attractive, because their jobs have only gotten harder and more complex in the last forty years. US taxpayer revolts, kicking off in California and spreading to other states since the 1970s, have been hard on public services like education (Archibald and Feldman 2006). This situation has only grown more dire since the 2008 financial crisis and the austerity implemented in its wake, particularly for public schools and public libraries—the two primary sites of investigation here.

While library visits surged with the unemployment rate, it took a decade for public library expenditures to return to their prerecession levels (Reid 2017). Three hundred thousand public school employees lost their jobs during the recession, and funding cuts exacerbated the inequality, growing for years, between rich and poor school districts—which also map well to majority-White and majority-non-White districts, respectively (Evans, Schwab, and Wagner 2019). Decisions about what to cut and how much to cut are of course always political decisions. US cities are particularly vulnerable to austerity measures implemented by state governments hundreds of miles away, state governments that may be hostile to cities' unionized workers, non-white residents, and left-liberal voters (Peck 2012).

And of course, when budgets are tight, schools and libraries must do more with less. Teachers become therapists. Librarians become social workers. Their superiors must find a pitch that appeals to politicians and donors. In these moments, the examples set by the startups down the street, not to mention the stories repeated by presidents and philanthropists, become powerful sources of inspiration.

The stress of a deeply unequal political economy makes places that do the teaching and training more receptive to the access doctrine. Inequality is a feature of a capitalist economy, not a bug, and the access doctrine makes this inequality sensible and navigable. It explains why there is such a gulf between rich and poor, how the poor can find security, and what help they need to get there—all without disrupting the basic shape of these unequal social relations. This is a positive, creative role that cannot be dismissed as mere propaganda.

At the most basic level, the access doctrine helps an overwhelmed high school teacher manage their classroom because they have learned that the smartphone is a social distraction, while the laptop is a professional tool for education, and they should discipline their students accordingly. Places like schools and libraries that provide tools and skills gain purpose through the story of the digital divide and sister crises like the skills gap or the STEM gap. They accrue legitimacy and resources by restructuring themselves and their work on these terms. The hope that personal computing, the internet, and the skills to use them will power social mobility is the cultural glue holding a deeply unequal information economy together.[4] That power resists refutation.

This hope in the power of technology to defeat poverty is first produced, and the concomitant crisis declared, by state institutions in the 1990s seeking to manage the national transition to an information economy and the persistent poverty within it. It is then repeated at the local level by public service organizations who are themselves under threat because it lends them purpose, clarity, and legitimacy. In the process, places like schools and libraries began to look more like tech startups.

I call this process of organizational restructuring prompted by the access doctrine *bootstrapping*: public service organizations, like schools and libraries, are overwhelmed by the scale of the problems facing them and find their resources and legitimacy under threat, so they turn toward technology provision and skills-training programs because these garner economic and political support and make the problems they face more manageable.[5] This is the focus of the ethnographic fieldwork that makes up the bulk of this book. Because bootstrapping empowers public service organizations that manage the problem of poverty, it is a never-ending process. Any part of the organization's identity, operations, and personnel are subject to revision. Paradoxically, it is precisely because the access doctrine presents such an urgent problem for these organizations that they will continue with new and different experiments in technology provision or skills training— even if those experiments do not benefit poor people. Those experiments serve the organization, even if they marginalize the people the organization serves. Bootstrapping becomes not just a series of changes but a new institutional culture, in conflict with older public service cultures.

When schools and libraries bootstrap, they are often inspired by the way technology startups pivot to new growth models. Although ultimately, they cannot pivot like startups because they have different goals, stakeholders, revenue streams, and responsibilities—which is why I mark their organizational restructuring with a different phrase. Take the W. E. B. Du Bois

Public Charter High School, the focus of chapter 4, as an example. Individual teachers worked hard to invest everything they did with the school's racial justice values—avoiding, for example, the "no-excuses" disciplinary measures common to other charters—but the school's high standards and the high stakes attached to them meant everything was subject to revision, including those racial justice values.

Models from the tech sector were everywhere, encouraging a culture of constant experimentation. Like other charter schools, Du Bois received regular funding from philanthropies that supported new technologies and technological skills for students, and training for teachers and administrators that encouraged them to think like entrepreneurs. Members of the leadership team had previously spun out their own startup, and while I was there, Du Bois piloted a student data management platform called School-Force, modeled on the Salesforce customer relationship management platform used at startups like InCrowd, the focus of chapter 2.

But the school faced a complex set of problems it was fundamentally unequipped to solve: child poverty, homelessness, gentrification, police violence, broken labor markets. And law and convention limited the school's flexibility in terms of when and how it admitted students or what benchmarks it had to reach to keep its charter. Still, the urgency of the access doctrine meant that there was no endpoint to organizational restructuring, no failure that could cause reconsideration of its mission. The school needed to keep transforming itself in order to transform its students. And as in MLK, the people the organization was built to serve were often sidelined in this process because bootstrapping kept the organization alive, securing much-needed resources and legitimacy.

Bootstrapping institutions reproduce the access doctrine and thus the idea of different sides of a digital divide, with one side in need of the other's help, as part of their general task of reproducing people for capitalism. This is the work Marxist feminists term *social reproduction*. It is work that occurs every day, both for "free" in the home and for a wage in spaces that may not produce a profit themselves (schools, hospitals, churches). Without such work laborers would hardly remain alive, much less able to work, socialize, consume, and generally navigate the world around them. What Katz (2001) describes as the "reproduction of the labor force at a certain (and fluid) level of differentiation and expertise" requires a whole geography of sustenance, knitted together by state institutions and cultural forms, with the terms of reproduction set as a result of social struggle between communities and global capital. Capital requires these spaces to raise, train, and support workers but, because they do not produce a profit, continually seeks to disinvest from them. In startups, schools, and libraries, I reveal a contemporary struggle over whether and how people of all sorts should be transformed into tech entrepreneurs.

This dynamic has held for centuries, changing with the mode of production. Capital has always needed unwaged housework or waged schoolwork as much as it needed waged factory work or unwaged plantation work, and it is within this sphere that different social roles are created and reproduced. Capitalism's birth centuries ago required the accumulation of differences among the working class so that, among other changes, social life and social roles were violently divided into private and public spheres (Dalla Costa and James 1975; Federici 2004; Fortunati 1995; Jackson 1940; Vogel 2013). Wage labor was undertaken by men in public and the care that prepared and repaired that laborer for work was done by women in the home; although women and children often continued to contribute, for much cheaper rates, at the factory or for free at home, and the lines drawn for White women were applied differently to Black women. In this model, the unpaid care work of women—whether at home or in the acts of care required to make any workplace run—and the sexist oppression of women is not outside or parallel to production for accumulation. Instead, this historical, never natural separation is crucial to the maintenance of capitalism as a system for the production of surplus value.

In the nineteenth and twentieth centuries, the working classes were largely divorced from the means of social reproduction and forced to rely on either the market, for commodities like food, or on racially segregated state institutions, for services like education (Mohandesi and Teitelman 2017). In the postwar era, an uneasy labor-capital alliance forced the state to take up a greater burden of social reproduction, creating, in an uneven manner across capitalist democracies, public rights to things like education, healthcare, and old-age pensions.

In the wake of the economic dislocations of the 1970s, the neoliberal state sloughed off these tasks and moved them into the market, where women of color frequently pick up the work of care that supports White professional women's labor force participation, or back into the home for those who cannot afford to pay (Bakker 2007; Bezanson 2006; Parrenas 2012). The

2008 financial crisis accelerated this dynamic. State disinvestment from schools and libraries, largely staffed by women, placed increasing stress on these institutions. This stress leads them to bootstrap. Because these are the institutions that teach us how to make a living, these changes to schools and libraries reverberate across the rest of society. As institutions of social reproduction restructure around a new mission for their own sakes, they teach everyone they touch that their future depends on securing the right tools and skills.

I attend to the labor of social reproduction throughout this book: the undervalued work of care, often women's work, that keeps people and institutions alive day to day and is so often left behind or forgotten in the masculinist drive for constant technological innovation. Social reproduction is not just a mechanical process of keeping people alive or training them in certain skills but, as Katz notes, the daily recreation of a set of cultural dispositions, a particular way of imagining the world of work, how it works, and what you are supposed to get out of it. It is the space in which we learn what the good life is and how to get there.

Within institutions of social reproduction, a culture of hope can bridge the gap between the inequity and conflict that surrounds them today and the justice they plan to enact tomorrow (Lashaw 2008). Hope can justify present disparities, or least mollify the people involved, by mobilizing a collective effort to address them. Like most US cities, DC is increasingly stratified by race and class and in the middle of major changes to its local labor market—all of which places great stress on its institutions of social reproduction. It is no surprise the access doctrine has taken root there.

The Digital District

It is easy to see the access doctrine's appeal in contemporary DC: a technological and professional path beyond a long history of labor market segregation that today unevenly distributes the spoils of the information economy. This is particularly true during the period studied here: the recovery from the 2008 recession. Approximately seven hundred thousand people live in DC. Recent gentrification has reduced its long-standing Black majority to a plurality. The city government has long hoped to lessen the local labor market's reliance on the federal government, the buildings and workers of which largely do not contribute to the city's tax base. These political-economic dynamics provided fertile soil for the access doctrine. A brief survey of the city is necessary to understand why.

DC is divided into eight wards. Wards 7 and 8 are east of the Anacostia River, majority Black, and long bereft of the jobs and services available in majority-White wards. This segregation accompanies one of the nation's most visible communities of Black professionals, historically concentrated in, but increasingly pushed out of, DC neighborhoods like Hillcrest (in Ward 7) and LeDroit Park (in Ward 1), as well as Prince George's County, Maryland, which borders the eastern half of the city along Wards 5, 7, and 8. As the site of the federal government, DC was, for most of its history, governed by Congress without any representation in Congress—a dynamic represented on its license plates with the slogan "Taxation Without Representation." In 1973, DC was granted home rule by Congress and began electing its own mayors and city councilmembers—although Congress still retained the power to block any law passed by the DC Council.

These may seem like unique circumstances, but just as most Washingtonians are regular working people, not politicians, these political dynamics are not qualitatively distinct from other cities. Rather, they are outsized manifestations of the city versus state conflicts that characterize most majority-non-White US metropolises—with Congress standing in for statehouses elsewhere.

Revenue has always been a problem for a city whose major employer and property owner, the federal government, is virtually recession-proof but does not pay taxes on its property, which makes up 50 percent of property in the city. The District is unable to levy commuter taxes against the 60 percent of its daily workforce that arrives from Maryland and Virginia and is ultimately able to tax only 34 percent of the income earned in its workplaces. This creates a structural imbalance in the annual budget, one that was estimated in a 2004 audit to be between $470 million and $1.143 billion (Office of the Chief Financial Officer 2004; for a broader overview, see Hyra and Prince 2016).

Decimated by White flight in the 1960s and 1970s, the 1980s and 1990s saw a commercial property boom in the city that never quite managed to reach its citizens, partly because many of the land sales either sold municipal tracts far below their worth or drastically discounted the price in anticipation of future sales tax revenues (Sherwood and Jaffe 2014).[6] The city's population reached a post-WWII low of 572,000 in 2000 as members of the Black middle class able to move did so (Bowser, De Witt, and Lee 2015).

The story began to change in 2008, as the worst economic downturn since the Depression gripped the majority of the country but largely missed DC and the surrounding suburbs. The federal government and associated contractors and nonprofits provided a base for this resilience, but the influx of new professionals—DC added nearly sixty-nine thousand people between 2008 and 2013 (Bowser, De Witt, and Lee 2015)—expanded beyond that to those working in education, hospitality, law, and technology (Dani 2013; Tavernise 2011). Although the private sector, including thousands of contractors who depend on the federal government, has always employed *more* people in DC than the public sector in the absolute, the 2008 recession marked a severe divergence in their relative growth rates, such that the overwhelming majority of job growth in DC after 2008 was in the private sector: 66,100 net new private sector jobs between January 2009 and March 2016 versus 4,000 of the same in the public sector (DMPED 2016).

The influx of new residents was raced and classed. It continued a trend begun under Anthony Williams's tenure as mayor at the turn of the millennium, in which the majority of in-migrants were higher-income Whites and the majority of out-migrants were lower-income Blacks—though higher-income Black households were more likely to move out than higher-income White households (Sturtevant 2014). New housing construction in this period was overwhelmingly concentrated in high-cost condominiums, and between 2002 and 2013, DC lost half of its affordable housing stock (Rivers 2015). By 2011, Chocolate City, for the first time in fifty-one years, was no longer majority Black (Tavernise 2011).

Gains from the tide of high-wage White migration were unevenly distributed. Postrecession job growth was concentrated in low-wage sectors alongside some increases in high-wage sectors, with the middle largely dropping out. While middle- and high-wage jobs saw their pay grow between 2007 and 2013, low-wage earners in DC saw theirs shrink, and long-term unemployment for those with only a high school diploma drastically increased (Lazere and Guzman 2015). Median Black and Latinx wages rose ninety and thirty cents per hour, respectively, in the same period, while median White wages rose three dollars per hour. In 2015, DC had the highest Black unemployment rate in the nation at 13.6 percent—or 5.7 times the White unemployment rate (Wilson 2015).

In 2015, as my fieldwork ended, the top 10 percent of income earn-ers in DC today made six times the bottom 10 percent, the highest dis-parity of any "state" (Giachetti 2015). This gulf emerged as the middle fell out of the local labor market after 2008, and especially after federal deficit-reduction measures began to take effect in 2011. A flood of young professionals entered the city and their wages rose, while low-wage service workers' wages stagnated and the federal government—the employer but-tressing the Black middle class in DC and its suburbs—shrank, in part by automating thousands of mid-level clerical positions dominated by Black women (Khimm 2014; Rein 2014). The uneven recovery did little to dispel the long-held conspiracy theory—"The Plan"—that White Washingtonians have been plotting to regain control of the municipal government and the city at large since home rule was granted in 1973 (Castaneda 2020).

The municipal government heavily courted these new, White, digital professionals and the firms employing them as the future of the city and as a way to create jobs and garner tax revenue without relying on the fed-eral government. They seemed to represent a cohort whose digital skills, information-processing jobs, and geographically mobile careers would help separate the city from the old, broken industrial economy and a federal bureaucracy that was historically unresponsive and unreliable with respect to the needs of the city around it. Development then, for individuals, com-munities, and the whole region, depended on making those who had been left out of the information economy more like those on the leading edge of it. Their tools, skills, and habits had to spread, or DC would get left out. This is what seems to drive "The Internet: Your Future Depends on It." This crisis and the proposed solutions to it are surely familiar to other cities in the United States attempting to create local jobs and grow the tax base, to reduce the reliance on commuters.

These political-economic dynamics place great pressure on schools and libraries and on the people who work in them or rely on them. Individuals reproduce the access doctrine as part of their jobs or as a means of securing recognition from institutions of social reproduction. Organizations repro-duce the access doctrine to secure resources and legitimacy.

This mode of neoliberal urban development pursued in DC and similar cities is the structural context for these social dynamics, pressuring orga-nizations to bootstrap. The municipal politicians and corporate leaders who manage urban economic development are themselves constrained by the movement of global capital (Tochterman 2012). For our purposes, it is important to mark out three tendencies of neoliberal urban development; the first two have been well-studied by urban geographers in particular, the third less so. They combine to create an urban political economy based on hope in technology and technologists.

First, largely White entrepreneurs are recruited from outside the city to relocate to it and invest in it. Second, the geography of the city is remade to entice, support, and reproduce these mobile professionals and their life-styles, through the management of the process commonly described as gentrification (e.g., N. Smith 2002, 2005). And third, urban institutions of social reproduction are rebuilt in the image of the firms that will move to the city, so that the people enmeshed in these institutions will emerge from them as entrepreneurs able to work in those newly arrived firms or ones like them—or start their own. Providence comes from innovative outsiders and/or drastic change to who the natives are and how they live and work. Internal solutions to these dire straits that might allow locals to build a regional economy independent of, or at least a stronger bargainer against, global capital are largely rejected on the basis that they would frighten capital away. A host of political, economic, and cultural processes drive these phenomena; gentrification is occurring in different ways in different cities across the world. But the access doctrine is especially important insofar as local instances of neoliberal urban development rely on incentivizing tech companies to relocate, supporting private development that caters to tech workers' lifestyles, and rebuilding public institutions in tech's image.

Much of this hopeful shift in urban governance, where local providence depends on wealthy White outsiders, is captured in Harvey's (1989) description of urban fortunes following the neoliberal revolution of the 1970s. This transition was marked mostly sharply by the near-bankruptcy of New York City in 1975, and then by a spate of downtown revitalization projects meant to drive White, monied consumers to shop there and producer services firms to relocate there, while heavily policing the Black and Latinx working and workless poor. Examples of such projects might include the Grand Central Partnership in New York, the Baltimore Inner Harbor, or, in DC, the MCI Center (later the Verizon Center, now Capital One Arena) and the Gallery Place development around it. Harvey names this trauma of urban abandonment followed by hopeful place-making as the shift from urban managerialism to urban entrepreneurialism.

Managerial modes of urban governance understood cities during and after World War II as outposts of the Keynesian welfare state. They were sites of industry with a diverse tax base and social services to match; many even engaged in direct job creation outside the municipal bureaucracy. The state-supported flight of the White middle class out of cities and the subsequent flight of industrial capital to the suburbs, the South, or overseas irrevocably changed this picture. Urban entrepreneurialism hangs a city's fortunes on remaking its downtown as a site for consumption and entertainment on the one hand and the headquarters for firms engaged in information-processing tasks on the other.

There were two primary results. First, those who lived in the city were confronted by a bifurcated labor market between relatively well-paid office workers and relatively lower-paid service workers who prepared the former's meals, clothes, and entertainment. This divide is often marked by race and immigration status. Second, those who ran the city found themselves no longer cooperating with other cities in different parts of the Keynesian industrial capitalist network (e.g., working at different parts of the national value chain or for different arms of the national state), but competing with other cities to create conditions favorable to increasingly mobile global capital.

This is the political-economic context in which the access doctrine takes root, not just as a strategy for individuals or organizations but as a philosophy of urban development that relies on upgrading the city and its citizens. The scope of this process is broader than the shift from managerialism to entrepreneurialism as described by Harvey. Cities like DC are certainly rebuilt to beat out peers and attract outside investment that will revitalize housing, labor, and entertainment markets. But Harvey's story is incomplete without an accounting of how the rest of the city, outside the offices of mayors or CEOs, experiences these changes, is taught to participate in them, and consents to them—or not. We must learn to hope in the power of technology and technologists to transform our cities.[7]

Hope here signals not just the goals of planners desperate for outside workers and outside capital, but the crucial role that reproducing that hope across the city plays in maintaining this system. Even and especially if that hope does not fit the lives of every person in the city, or every institution that cares for it. New strategies of capital accumulation require new modes of social reproduction. For city dwellers, changes in the local economy are often most keenly felt in this sphere: where they can afford to live, whether the neighborhood school stays open, how much public transportation costs. This is where urban citizens are enlisted in the project of building a new, entrepreneurial city.

Overview of the Book

To understand these urban and organizational dynamics, we must first understand their historical context. Chapter 1 reveals how the access doctrine turned the dire straits of persistent poverty in the neoliberal era into a hopeful, technological opportunity. The invention of the digital divide in the Clinton era provided a sense of mission and urgency to what otherwise appeared to be mere technology provision or skills training. This is where the charge to bootstrap comes from, and it is why the survival of public service organizations depends on their willingness to commit to the project. I show that the access doctrine was a project borne of a particular neoliberal coalition with particular race and class politics, popularly known as the Atari Democrats.

The remainder of the book explores those organizations mobilized by the access doctrine, which seek to close the digital divide. It is only at the level of everyday organizational life that we can see how the problem of poverty becomes a problem of technology—for whom, why, and with what results—and it is only by moving between organizations that we can understand the social and political networks that convince these places to bootstrap, rapidly remaking themselves to solve the crisis of the digital divide.

Where this introduction and chapter 1 focus on the structural scale of large political-economic changes, chapters 2, 3, 4, and 5 focus on mesoscale organizational life.[8] Beyond the particular dramas of InCrowd, MLK, and Du Bois, the bootstrapping process also provoked extensive debate over the nature and purpose of startups, schools, and libraries as social institutions.[9] The conclusion reintegrates these different scales into a larger theory of the political-economic role of the access doctrine and how a new coalition could challenge it.

Where chapter 1 focuses on the access doctrine's birth at the level of national politics, chapter 2 zooms in on startups, particularly one called InCrowd. Here we see the reproduction of the access doctrine within individual subjectivities and organizational philosophies through a case study of the "right" side of the digital divide. I explore the importance of the pivot to this culture: a moment of fundamental redirection in a firm's purpose and operations. Over and above any particular product or skill, the ability to pivot is what grants the disparate corners of tech coherence as a sector and drives employee buy-in to a workstyle dominated by uncertainty.

Chapters 3 and 4 focus on libraries and schools, respectively, and their attempts to bootstrap. They tell the story of urban institutions remaking themselves to look more like startups and remaking the people within them to look more like internet entrepreneurs. It is here that we see the bootstrapping process unfold and compare it with the role of the pivot in startup culture.[10] This is a contested process. Older institutional cultures based in public service still motivate these organizations and their personnel, provoking resistance to organizational restructuring. And library patrons and high school students engage in their own place-making projects within these institutions, but on terms that do not conform to the terms of debates between White professionals over what role technology should play in helping the poor.

Chapter 3 focuses on the daily life of the computer labs in DC's MLK central library branch, one of the last few safe public spaces for homeless Washingtonians. In their day-to-day work, the access doctrine provided librarians with a way to manage social pressures they were unequipped to solve and decide who is a "good" or "bad" library patron. Over the longer term, the library itself was remade into more and more of a skills-training center to secure a much-overdue renovation and justify its place in a changing city.

Chapter 4 explores a similar conflict between engrained values and a new mission in an entrepreneurial charter school. We enter W. E. B Du Bois Public Charter High School (a pseudonym), as its first senior class—the prototypes for Du Bois's educational philosophy and innovative technology programs—begins its final year, school-issued laptops in hand. Up to this point, faculty had largely been able to reconcile the conflict between the school's broad, racial justice values and its narrower mission: social mobility measured in test scores. But administrators grew alarmed as the seniors' performance data rolled in, and as graduation approaches the school's digital infrastructure was used to subvert its values to its mission. The imbalance of power in the school's governance forces it to bootstrap.

Chapter 5 compares the school, library, and startup in order to explain why bootstrapping happens, why MLK and Du Bois begin to look more like InCrowd. The hope of the new MLK central library branch was invested in the glass-walled collaborative workspaces and makerspaces used by library visitors, in contrast to the quiet rows of the computer lab used by homeless patrons every day. Du Bois engaged in a constant series of data-prompted revisions to its mission and methods, attempting to pivot like a startup. I examine the threads that tie these sites together across the city—especially philanthropies and professional networks—and show how the idea of skills training and technology provision becomes sensible, logical, and urgent.

Hope powers change, but often fails in the attempt. Because of the outside stakeholders and historical constraints on the mission of these public institutions, they can never fully live up to the example set by their more nimble, private peers. The conclusion examines the failures of the bootstrapping model alongside the glimpses of political power we can see in how homeless library patrons or working-class high school students build their own places of support and relief within the larger organization. By understanding how this process fails and how people survive within it, we can begin to see political alternatives, potential coalitions that can organize at the point of social reproduction.

Methodology

I conducted thousands of hours of ethnographic fieldwork between 2012 and 2015, in organizations and with people across DC, to examine how the access doctrine circulated among different people and organizations and motivated changes in their work and identity (Gowan 2010; Marcus 1995). I supplemented my participant-observation with interviews of fifty-five people, many repeated across several years. This proved especially useful for engaging startup workers and founders who were much less tied to any single workspace than teachers, librarians, patrons, or students (Hannerz 2003). I also amassed a collection of texts, ranging from political flyers to promotional materials to syllabi. Fieldwork began in the spring of 2012 and concluded in the summer of 2015, though I returned for follow-up interviews and site visits several times over the next three years—particularly to check on the library's closure.

In my analysis, I approach these different data streams as ingredients in an *institutional ethnography* that is less about the story of individual people and more about the social relations within and between these different places (DeVault 2006; D. Smith 2005). I focus on mesoscale organizational life in part because that is, empirically, where the action was and, politically, because urban ethnography in the United States has a long history of scandalously objectifying the lifeways of the racialized poor through a focus on individual behaviors that then circulate beyond their original context (Gowan 2009, 2010; Rios 2015; Wilse 2015). I hope to avoid this dynamic by keeping the focus on, for example, the space of library, rather than the behaviors of individual homeless patrons in it.

I fit into each place differently. At InCrowd, I was a peer professional. In both MLK and Du Bois, there was an obvious power imbalance between myself and the people being served by the institution (students, patrons) and a just as obvious demographic and professional symmetry between me and the helping professionals serving them (librarians, teachers). This was visible in my dress, my Apple laptop and iPhone, my education, and my whiteness. That I was a former social worker helped me hold conversations about the difficulties of passing through the system, especially for patrons but also for several students who were in and out of shelters, but this was just another inequality; I have never slept rough. Insofar as I was not empowered to discipline anyone[11] and was interested in learning about all the things one should *not* do in that space, I built rapport as a friendly outsider. Helping with homework or job applications helped build those relationships while leaving the power imbalance intact. A fair trade for entrée, but just as meaningful were long, meandering conversations about superhero movies or the hip-hop soap opera *Empire*.

Early thoughts on particular chapters were shared with key informants, who also gave feedback on chapter drafts over the years this project developed. Greater care was taken with portions of the manuscript describing students and homeless patrons, both in terms of seeking out their feedback and anonymizing their names and locations within the institution (i.e., people's appearances, their usual seats). Despite this collaboration, this is by no means a collaborative ethnography in which informants coauthored the story (Lassiter 2005). The power relation between ethnographer and informant remained intact throughout. While I strive to treat their stories with respect, I have no pretensions regarding this book's ability to "give back to" or "speak for" my informants, particularly students and patrons (TallBear 2014). Academics' dissertations and first books have, by necessity, a limited reach, a specific vocabulary, and a lengthy incubation period—all of which circumscribes any political ambitions.

This is not cynicism. It is a recognition of my informants' exceptional political literacy. They require no social scientific ventriloquism from me. Du Bois's students organized a walkout against police violence while I was there. Some librarians lived in fear of the email campaigns that homeless patrons had launched against specific branches of the municipal government. Because of this, I do not consider my role here to be giving back to a specific culture, much less speaking for it. I merely hope to tell these stories fairly and frankly, explaining a process that has connected us all, in different ways and in different social locations, in the map of hope and inequality.

1 Discovering the Divide: Technology and Poverty in the New Economy

In September 2014, Connect.DC, an internet access initiative from DC's Office of the Chief Technology Officer (OCTO), hosted a one-day symposium titled "Bridging the Digital Divide in the District of Columbia." Attendees included representatives from every major social service agency, nonprofits from across the city, the startup community, and the DC Chamber of Commerce. We were housed in the first floor of the *Washington Post*'s offices, down the street from the 1776 startup accelerator that had rapidly become the center of the District's tech scene.

On the surface, this was an occasion to review the outcomes from four large internet infrastructure grants from the National Telecommunications and Information Administration (NTIA) the city had received as part of the recovery effort during the 2008 recession. But the stakes were a good deal higher. The city—at least the largely white portion west of the Anacostia River—had done quite well during the recession. But Black unemployment was still stubbornly high, and those new high-wage migrants were driving up rents. It was clear to everyone in the room that even though 150,000 DC households still lacked an internet connection, digital inclusion meant far more than a broadband hookup. Advertising the event, and echoing Al Gore from two decades earlier, Connect.DC asked, "How can we close the gap between the 'haves' and 'have nots' in the District of Columbia?" Moderator Henry Wingo, from the Chamber of Commerce, called it "a gulf between worlds." A year later, he would sue to keep a referendum on a fifteen-dollar minimum wage off the ballot.

Digital inclusion was a priority not only because certain jobs or services might be out of reach but because, in the words of Michelle Fox, of nonprofit Code for Progress, we were "leaving talent on the table." Richard Reyes-Gavilan, new executive director of DC Public Library, said the same, emphasizing that libraries were today more important than ever as premier sites of "human capital development." He said that he had come to town to rebuild the MLK central library branch, making it a friendly place for the community and the burgeoning startup scene. Patrick Gusman, cofounder of Startup Middle School, bemoaned the lack of technical curricula in DC schools and prescribed computer science as a mandatory language capstone course in which students would "use a form of technology to solve a social problem." Wingo himself compared the urgency of this conference to that of the civil rights movement.

This idea that a lack of technology and technological skills is what is holding back poor people and poor regions is relatively new, historically speaking. This chapter seeks its source, turning back the clock to uncover the roots of the access doctrine—the political common sense that held that poverty can be solved with the right digital tools and skills. The access doctrine empowers Reyes-Gavilan to bootstrap the MLK Library, remaking not just how it looks but also what it is and who it is for. I find the source of this political common sense in the politics of the Clinton administration as the internet commercialized and it identified the problem of the digital divide. The access doctrine undergirds the digital divide framework, but that work continues even as *digital divide* fades as a term of art in policy, to be replaced by ideas like *digital inclusion*. The access doctrine's hope keeps appearing in places like the OCTO event or the Connect.DC posters—"The Internet: Your Future Depends on It."

By focusing on the relationship between technology policy and poverty policy, we can see how the problem of access becomes a problem in the first place, in terms everyone understands, with solutions everyone agrees on: public-private partnerships for access extension, the demand that everyone must learn to code, and the wide embrace of a startup mindset that turns welfare state bureaucracies into nimble, twenty-first-century service providers. The creation of the problem is a project of a particular race and class coalition, known at the time as the Atari Democrats, who are still visible at the OCTO events: largely White professionals at the base, led by business leaders from the tech sector and closely linked industries that require state support for their global ambitions.

The Clinton administration's first report on stratified internet access in the United States, what it would eventually call the digital divide, argued, "While a standard telephone line can be an individual's pathway to the riches of the Information Age, a personal computer and modem are rapidly becoming the keys to the vault" (NTIA 1995). What is left out of this story is how the vault became locked in the first place. This includes, beginning in the 1970s, increased automation or outsourcing of industrial production, stagnant real wages, increasing healthcare and higher education costs relative to inflation, rollbacks of federal poverty relief programs, and the massive expansion of the carceral state in poor communities (Edelman 2013).

The question is not whether the inequalities identified under the rubric of the digital divide were real. They were. In 1995, urban cores and isolated rural areas really did lag behind wealthier suburbs in internet access, and Black and Native communities really were less well-served by internet service providers than White ones. The question instead is why inequality was described in this way and how it connects to other pieces of contemporary poverty policy. Three pieces of the access doctrine appear in succession: a crisis of national competitiveness is declared, it is then defined in human capital terms, and finally resolved through deregulation of telecommunications markets and targeted public-private partnerships. Exploring these steps provides a new entry point for understanding the post-1970s dismantling of the Keynesian political consensus, its reconstruction as neoliberalism, and the discursive role information technology played in this shift. The access doctrine has deep roots, so first we must briefly review the success of the neoliberal political project in redefining the structural problems of economic dislocation as individual deficiencies of skill, a process that began in the 1970s.

Technology Policy as Poverty Policy in the Neoliberal Era

There is a core contradiction in contemporary US poverty policy. Since the 1970s, political institutions and policymakers have advanced two seemingly opposed positions. On the one hand, the neoliberal state must offer promise: with the right skills, the global labor market becomes a space of unlimited potential where anyone can become an entrepreneur. On the other hand, the neoliberal state must threaten punishment: anyone who steps out of line will, at best, have their state support revoked or, at worst, be incarcerated.

The access doctrine emerges from the Clinton administration in the 1990s, as it oversaw the commercialization of the internet and named the problem of the digital divide. It resolves the contradiction between promise and punishment within American poverty policy by presenting a relatively simple recipe for economic security in insecure times: digital tools and skills lead to good jobs. The internet, the story goes, unlocks the fetters of geography and identity. A series of MCI commercials, for example, promised that on the internet there is no race or gender, and no "there," only "here" (Nakamura 2000). Promise exists anywhere the internet does, and so the blame for poverty must fall on those who cannot act on that promise. Within the access doctrine, race in particular appears as a stubborn remainder of an older political-economic order. It is a problem that must be either resolved with particular upgrades or, if that doesn't work, contained by the carceral wing of the state.

To see this common sense emerge, we must understand attempts to bridge the digital divide or teach people how to code not just as *technology* policies, but as *poverty* policies, working in tandem with attempts to reform unemployment insurance, job training, criminal justice, and so on. In this way, access initiatives are better understood as one component of a new script for the state that reduces or obscures the state's ability to guard against periodic economic crises and instead highlights its role as a guarantor of competition for its citizens, themselves circumscribed as bundles of human capital entering the market to contribute to national economic fitness.

The access doctrine was not invented out of whole cloth. Technological hope has a long American history. In the 1960s, welfare-state liberals often imagined technology not just as an engine for individual social mobility, but a solution to the urban uprisings of Black Americans who had been kept out of midcentury industrial prosperity. At this stage, the connections between technology policy and poverty policy were clear. RAND Corporation researchers, for example, argued that cable could "provide a possible antidote to the isolation of 'ghetto residents'" (Light 2001, 718). These attempts to "keep the peace" with the working and workless poor are emblematic of the Keynesian political consensus (Gilmore 1999). Across Western democracies in the postwar era, industrial capital constructed a series of national safety nets for itself by consenting to the creation of safety nets for citizens that included such benefits as old-age pensions, unemployment and disability insurance, large infrastructure projects, subsidized housing, expanded higher education, and more. Such benefits were hard won by the working class but were divided unevenly across that class by race, gender, and citizenship (Katznelson 2013). This consensus kept the peace with industrial labor in particular, guaranteed high rates of consumption, and ensured a few decades of steady growth.

But that growth did not reach everyone. Increasing automation and off-shoring (even just to nonunion Sunbelt states) in the 1960s hit Black communities first and hardest, driving up unemployment rates in the central city areas to which Black families had migrated earlier in the century (Sugrue 2014). Some reformers saw computer training as a solution to these problems of Black poverty and social isolation. Alan Bekelman, a former high school teacher who despaired for his students' job prospects, pitched the idea to superiors in the Commerce Department. His Scientific-Technical Employment Program (STEP) trained two hundred students in programming every year as part the larger Youth Opportunity Campaign summer jobs program (Loftus 1967). Beneath the "Companies Aid Computer School" headline, a July 1968 *New York Times* article told the story of twenty-year-old "youth from Harlem" Van Sloan and 140 compatriots "learning the skills of key-punching, computer operations and programming at the Middle West Side Data Processing School, an antipoverty training center." Twenty-one-year-old Sherry Barnes explained that she quit her day job to enroll because "the computer is a passport to a better job and a guarantee to a sure future, if you are smart enough to take advantage of it" (Smith 1968).[1]

While pieces of the access doctrine are certainly visible in the Great Society era, those pieces could not come together with the momentum necessary to really make things happen. That would have to wait for the neoliberal revolution. Beginning in the 1970s, revanchist political alliances sought new solutions to crises of profitability born of global economic conditions (e.g., stagflation, the oil crisis), constraints national governments had placed on global capital, and militant social movements the world over. Industrial unions' power receded as production was increasingly automated or moved beyond organized labor's reach, first to the periphery of the industrial core and then to the Global South (Silver 2003). The service sector jobs that replaced them had different workplace structures, energized employers supported by an increasingly antiworker state, and a more feminized and racialized set of workers who were both under threat and frequently ignored by traditional unions (Cobble 1991; Lopez 2010). This weakened labor's power, although there were of course notable exceptions to the trend, particularly in healthcare (Windham 2017).

As the mode of production shifted, policing powers were built up and redeployed. A new carceral state emerged in reaction to the power of mid-century social and labor movements and White fears over Black urban unrest (Berger 2013; Camp 2016). A prison-building boom followed, absorbing surpluses in land, labor, financial capital, and state capacity borne of the economic dislocations of the 1970s (Gilmore 2007).

This is the economic context that both nurtured and was nurtured by *neoliberalism*—a slippery term. I follow Wacquant's (2012) definition of neoliberalism as a political project wherein an activist state repurposes its institutions to define and enforce citizenship around market demands. This requires not a shrunken state, but a reengineered one: enhanced, redistributive bureaucratic functions for the upper classes alongside more paternalistic functions for the lower classes, and a massification and glorification of the penal system. The latter is especially important insofar as it shows that the neoliberal state is neither hands-off, focused primarily on deregulation, or miniaturized, with all its various capacities downsized together. The neoliberal state absorbs and redirects the state capacities created by the Great Society.

I diverge from Wacquant, and follow Gilmore (1999, 2007), in considering the prison a model neoliberal institution insofar as it does not only warehouse (particularly Black) people labeled *problems* but also teaches other people, whether or not they touch the prison system, and institutions the rules of the game for a new political economy. Schools and libraries are part of the same project and so express similar rules; they do not just react to political-economic shifts but produce and reproduce them (Sojoyner 2013; Willis 1981). Although they and their personnel often conflict in their day-to-day operations, these institutions of social reproduction are united in their efforts to develop and direct the capacities of people, places, and public funds toward the ends deemed most "productive." One set of institutions, studied here, offers opportunity; the other, lurking ever in the background of the lives of the urban poor, punishes those who do not follow the rules and accept the offer. In the 1990s, those rules became, to put it crudely: log on, train up—or else.

This approach took root in the Nixon era, but was cemented by Clinton and his promise to "end welfare as we know it." It was part of a new mode of social reproduction that remade people for a new labor market, even if they were currently outside it. Categorizing the deserving unemployed through unemployment insurance or disability insurance, for example,

implies a group of undeserving unemployed. Both categories require a definition of the ideal worker against which they can be compared (Piven 1998;
Piven and Cloward 2012). Both categories underwent massive revision in
the transition to neoliberal political rule.

Prior to the 1970s, in the period from Roosevelt's New Deal through
to Johnson's Great Society, welfare relief integrated with broader goals of
industrial policy: "full (male) employment, mass consumption, stabilized
social reproduction, and a particular pattern of labor segmentation" (Peck
2001, 46). After much struggle on the part of aid recipients, particularly
Black mothers, a general right to welfare was established by the 1960s, supporting some of the work of household social reproduction and acknowl-
edging the gendered terrain of the labor market that either excluded or
underpaid poor women (Nadasen 2002; Reese and Newcombe 2003).

These gains were short-lived. As the urban uprisings receded and US
industry stuttered under stagflation, Nixon framed both welfare recipients
and the institutions serving them as impediments to national economic
progress, local productivity, and individual morality: "The task of this government, the great task of our people, is to provide the training for work,
the incentive to work, the opportunity to work, and the reward for work"
(1969). The state no longer offered a safe harbor; instead it promised to teach
you how to sail. Macroeconomic solutions to poverty receded, replaced by
training solutions (Lafer 2002).

These skills-training initiatives, like Reagan's Job Training Partnership
Act, occur within a more general punitive turn in poverty policy. This
included a sharp reduction in federal housing services and, through budget-
ary maneuvers, increased eligibility restrictions for the Aid for Families with
Dependent Children (AFDC) program and increased funding for state workfare programs. Five hundred thousand people were removed from AFDC,
and states were encouraged to remove more through experiments with
workfare programs: "the imposition of a range of compulsory programs and
mandatory requirements for welfare recipients with a view to *enforcing work
while residualizing welfare*" (Peck 2001, 10; emphasis in original).

Reagan also started a trend of granting states waivers for welfare programs that revised traditional eligibility standards. Both Bush and Clinton
issued waivers liberally, as long as the proposed local revisions were cost-
neutral. This discouraged states from implementing more expensive, service-
oriented workfare programs that provided long-term, high-quality training

and counseling, along with other services like transportation or childcare. Cutting the rolls with stringent eligibility requirements was cheaper and easier (Peck 2001, 100). It would seem, then, that the punitive turn undercuts the hope of skills training: How can those on the fringes of the labor market train for the jobs of tomorrow if they cannot put food on the table today? It would take a different set of politicians to marry these two conflicting tendencies.

The conservative political advance of the 1970s and 1980s reframed the role of the state and the problem of poverty in the United States. It also proved immensely popular, and the GOP was rewarded with almost two decades of electoral supremacy in the White House and Congress. But a new generation of Democrats built on these policy lessons with a winning spin on the problem of poverty. They linked issues of skills and issues of dependency into one broader problem: a shortage of willing and able workers to staff the high-tech jobs of the future. This was a more hopeful vision than the one presented by Reagan—who focused on an impotent state and the cheaters trying to scam it—but the people and institutions identified as the problem remained the same: non-White, especially Black, poor and working-class people struggling on the fringes of the labor market, and the welfare state institutions serving them. The racist caricature of the Black "welfare queen" and the broken bureaucracy serving her was wielded by both parties (Hancock 2004; see also Fraser and Gordon 1994).

Clinton and Gore were allies in the Democratic Leadership Council (DLC) of the 1980s, moving the party rightward, away from New Deal social democracy, in order to reverse years of Republican electoral gains. On the campaign trail during the early 1990s recession, and once in office, they described their new generation of Democrats as superior economic managers, willing to make some Keynesian investments in human capital and export-oriented industries, opening borders to free trade, and focusing on deficit reduction (Ferguson 1995). The latter limited any potential stimulus that would counter the recession and showed that Clinton and Gore's opposition to Reagan and Bush was largely a matter of strategy, emerging from a different electoral base rather than a fundamental political disagreement. They were two different wings of the neoliberal project.

The "discovery" of the digital divide and the birth of the access doctrine cannot be analyzed without this context. The Clinton administration positioned its support for digital training programs for disabled Americans, for example, within a larger mission to "give work back to the American people" (Clinton 2000b), without ever endorsing direct stimulus or job creation. This included the effort "to end welfare as a way of life and make it a path to independence and dignity" (Clinton 1993), which resulted in the Personal Responsibility and Work Opportunity Act (PRWOA) of 1996. Clinton and his Congressional allies celebrated PRWOA, which replaced the AFDC poverty relief program with Temporary Assistance for Needy Families (TANF), for supplanting the American poor's entitlement culture with a work culture.

Funding for TANF was block-granted so that a countercyclical poverty policy became nearly impossible,[2] while limits were placed on recipients who were not working or who were unwed mothers or undocumented immigrants, on top of the five-year lifetime limits applied to everyone. No new training or job-creation programs were paired with these new restrictions (Wacquant 2009). A decade after PRWOA, an additional three hundred thousand children were living in poverty because of new restrictions on aid to households headed by single mothers (Trisi and Saenz 2020). In 1996, 4.7 million families were receiving cash assistance under TANF, by 2013, under AFDC, that number fell to 1.7 million—and never exceeded two million during the great recession (Shaefer, Edin, and Talbert 2015). Unsurprisingly, the percentage of nonelderly households in extreme poverty—surviving on less than two dollars per day—grew sharply in this period, from 1.7 percent in 1996 to 4.3 percent in 2011 (Shaefer and Edin 2013).

These policies punished those already hit hard by deindustrialization and the failure of federal poverty-relief measures to keep pace with inflation, as well as the shorter-term damage of the early 1990s recession. The punishment would continue. In 1994, as part of the Atari Democrats' tough-on-crime agenda, the Violent Crime and Law Enforcement Act created sixty new death penalty offenses, criminalized gang membership, ended Pell Grants for college education in prison, and funded almost one hundred thousand new police officers. Contemporary projections indicated that these changes, echoing similar state and local policies underway for decades, would result in the incarceration of at least 1.5 million new people and require nearly $351 billion—almost twenty times the 1994 AFDC budget—in new prison operation and construction funds (Duster 1995).

The early 1990s were not only the peak of the punitive turn in poverty policy but also a pivot point for the American political economy. It was the moment when Clinton and Gore pushed the country toward a hopeful New Economy[3] led by the tech sector. This is no coincidence. Gramsci (2000, 222–245) argues that in moments of economic transition, when the reins of power are up for grabs, political coalitions secure power partly through activist cultural policy that emphasizes sharp breaks with a denigrated past and new institutional directives fit for new economic demands. The early 1990s were one such moment. The access doctrine helped unite New Economy boosterism with a revanchist social state by telling a new story about what success meant and how it could be achieved. This story has three parts, each building on the last: a declaration of national economic emergency, a human capital measurement project, and a conclusion that the state could not solve a problem of this scale. The state would conduct triage and leave all but the most urgent cases to a deregulated telecommunications sector. In this process of naming, measuring, and attempting to solve the problem of the digital divide, the access doctrine was borne. That common sense had such political power that it would outlast the specific framework of the digital divide. These features—crisis, measurement, public-private partnerships for solutions—would then go on to appear in new frameworks such as digital inclusion or more recent calls to learn to code.

The Digital Divide as National Economic Crisis

The neoliberal substitution of skills training for job creation and poverty relief set the stage for the poverty policies of the information economy and the scripts for the institutions that would implement those policies. But early 1990s economic conditions raised the stakes, such that even before anyone spoke the words *digital divide*, it was understood to be a crisis. This is the first part of the access doctrine: a crisis of connectivity, which, because workers cannot connect to the global labor market, becomes a crisis of underutilized labor.

During the 1992 election campaign and throughout its first term, the Clinton administration argued that getting every American plugged into a National Information Infrastructure (NII) was a matter of economic survival. Investment in the fixed capital of fiber optics and the human capital of skilled professionals would cement victory over Soviet communism, end the early 1990s recession, and regain global economic dominance from

Germany and Japan. The problem of the digital divide and the solution of the access doctrine are rooted in this refigured economic nationalism. *Refigured* because while Clinton and Gore distinguished themselves from Reagan and Bush by endorsing some Keynesian stimulus on the campaign trail, most major stimulus plans were dropped in favor of deficit reduction after they took office—and the NII proposals that persisted were, compared to Roosevelt's rural electrification or Eisenhower's highways, relatively modest in scope (Ferguson 1995).

Clinton's 1994 budget, for example, requested from Congress $1.1 billion specifically for the National Information Infrastructure and $45.6 billion for more general training and vocational and adult education (largely grants to states and municipalities) meant to "add to the stock of human capital by developing a more skilled and productive labor force" (72). For context, $277 billion was requested in military spending and $2 billion for modernization and expansion of federal prisons—before the Violent Crime and Law Enforcement Act.

Raw numbers aside, the Clinton administration's plan to connect every American to the newly privatized internet was framed as an investment in national economic competitiveness. Within this first stage of the digital divide crisis, combating poverty is a problem not only of alleviating suffering in the present but also of making the correct investments in "information have-nots" so as to resolve current crises of underutilized labor, realize future capital growth, and achieve post–Cold War international economic hegemony.

In 1991, then-Senator Gore proposed the High Performance Computing Act to study how to upgrade NSFNET—the still limited, still research-dominated civilian internet—for commercial and consumer use. The bill apportioned $1.547 billion to the NSF to support new regional internet service providers and to build regional inter-networking points that would connect a privatized internet. During the 1992 campaign and immediately after taking office as vice president, Gore repeatedly posed NII buildout as a national economic emergency. Political opponents attacked this as undue state intervention. But Gore had spent years carefully negotiating this terrain, publicly connecting networked technologies to collective economic fitness, individual consumer choice, and democratic deliberation. He argued for his NII proposals in a 1991 issue of *Scientific American* alongside other early internet architects:

The unique way in which the US deals with information has been the real key to our success. Capitalism and representative democracy rely on the freedom of the individual, so these systems operate in a manner similar to the principle behind massively parallel computers. These computers process data not in one central unit but rather in tiny, less powerful units.

Capitalism works on the same principle. People who are free to buy and sell products or services according to their individual calculations of the costs and benefits of each choice process a relatively limited amount of information but do it quickly. When millions of individuals process information simultaneously, the aggregate result is incredibly accurate and efficient decisions. . . . Communism, by contrast, attempted to bring all the information to a large and powerful central processor, which collapsed when it was overwhelmed by ever more complex information. (151)

This conflation of different scales—infrastructure and individual, personal computing and national markets—was not just Atari Democrat spin, but an overarching regulatory regime emphasizing market competition as the primary political calculus and market citizenship as the primary political unit. Nor was the anti-Communism element simple cheerleading. Clinton and Gore (1993) positioned NII buildout and basic research into technologies of "commercial relevance" as the place to shift funds no longer required for Cold War militarization—even if the actual funds, both requested and disbursed, were paltry compared to what went into prisons or defense.

Because the internet would necessarily exceed the boundaries of the United States, it was also posed as an instrument of soft power—especially within newly capitalist, post-Soviet states (Gore 1994c)—to the benefit of US software producers that supported the Clinton-Gore campaign and depended on both liberalized trade and the dominance of English as the language of commerce (Ferguson 1995, 301). The administration took this economic nationalism so seriously that while running for reelection Gore accused Bob Dole of "unilateral disarmament" for threatening "to cut America's science and technology budget by one-third" (Clinton and Gore 1996b).

Internet infrastructure buildout was a crucial part of the administration's plan to upgrade the workforce for an information economy. This New Economy was based on transmitting and manipulating information but was not limited to software coding or computer manufacturing: it was post-sectoral. "Everyone will be in the bit business," Gore said (1994c). Within the Technology for America's Economic Growth policy initiative, released a month after Clinton and Gore took office, any gaps in connectivity were a blow to the nation's standing in the New Economy—to the point where "schools can themselves become high-performance workplaces" to train tomorrow's technologists (14).

"Because information means empowerment, the government has a duty to ensure that all Americans have access to the resources of the Information Age"—a duty that, in the administration's telling, Reagan and Bush had neglected (Department of Commerce 1993). But that duty did not demand traditional Keynesian public works responses. The provision of access was meant to create new markets or better position American exporters in existing ones—not provide market alternatives. Funding requests for infrastructure buildout were not particularly large, certainly not sufficient stimulus for the early 1990s recession: $600 million for the High-Performance Computing Act, $100 million per year for the NII (Department of Commerce 1993). The bulk of the cost for extending the commercial internet to every American would be borne by telecommunications firms incentivized by deregulation.

This mild Keynesianism was supported by an investment bloc of capital-intensive, export-oriented industries, especially high-technology companies that felt threatened by the rise of German and Japanese competitors. They needed the state to relax import tariffs for components, negotiate lower export tariffs abroad, educate a new generation of professionals, protect intellectual property, and provide at least the groundwork for an internationally competitive communications infrastructure (Ferguson 1995). Donors from this sector, including lifelong Republicans such as John Young of Hewlett-Packard and John Sculley of Apple, formed the Council on Competitiveness and provided pivotal funding and public support for the 1992 Clinton-Gore campaign (Sims 1992). Many of these elite donors were then recruited to the Advisory Council on the NII to advise the secretary of commerce on all matters internet.

This cemented the Clinton administration's links with high-technology companies and created a new business coalition to challenge the one that had backed the Bush and Reagan administrations, which had been largely based in oil, food, investment banking, and less-mechanized manufacturing sectors such as textiles (Cate 1994; Ferguson 1995). This was the C-suite version of Clinton and Gore's appeals to the Atari Democrats' electoral base: A new voting bloc identified with office park corridors like that of Route 128 in suburban Massachusetts and named after the popular Atari video-game system (Geismer 2014; Wayne 1982). As organized labor retreated, the

Democratic Party increasingly prioritized these largely White, highly educated professionals. In the primary campaigns of 1984 and 1988, these voters
were organized against the multiracial working-class coalition of Jesse Jackson's Rainbow Coalition.

Speaking at the 1997 Microsoft CEO Summit, Gore emphasized the competitive advantages his public-private infrastructure project had borne and
warned of isolationists and fiscal conservatives attempting to stymie his
efforts. He was clear that extending access to all Americans was a key part of
a rich and free capitalism. Whereas in the old economy "growth depended
largely on capital and labor [and] the task of policy makers was to keep those
factors of production in sync," in the New Economy the main assets were
ideas, "our core capacity as human beings," brought to market through the
internet.

The Digital Divide as Measurement Program

While the Atari Democrats framed their technological investments against
the "pure" laissez-faire of neoconservatives, their interpretation of poverty
in the New Economy as a national crisis of competitiveness—and the proposed definitions of and solutions to that crisis—was strikingly similar to
neoconservative ideas about crises in education. Secretary of Education Terrel Bell convinced Reagan to make education a conservative issue—and not
defund the agency Bell led—through 1983's *Nation at Risk* report. It framed
a decade of falling SAT scores, in an era when the pool of test-takers rapidly
expanded, as a "rising tide of mediocrity" that left students so deficient in
the skills needed in the global labor market that "if an unfriendly foreign
power had attempted to impose on America the mediocre educational performance that exists today, we might well have viewed it as an act of war"
(National Commission on Excellence in Education 1983).

Then governor Clinton had picked up this torch as chair of the 1989
National Education Summit, endorsing a national program of outcomes-
based standards, charter schools, and standardized testing (Scott 2011).
Schools here were positioned not as welfare state social supports but as
skills-training centers. With the federal deficit ever in mind, supporting
these skills-training centers demanded a program to measure the extent
of the skills gap and carefully target interventions. Standardized testing
was one manifestation of this measurement program. The mapping of the

digital divide was another, a project to identify gaps in internet access and digital literacy so as to best direct the public partnerships that would provide PCs, modems, and training.

After the Atari Democrats declared a crisis of national economic fitness, it had to be mapped so that appropriate interventions could be identified. The second piece of the access doctrine sought to measure the depths of the crisis. Digital divide policy framed problems of poverty as problems of performance; poor people were poor because of insufficient or underutilized human capital. The state needed to understand the nature of underskilled, uncompetitive labor and the locations where technological infrastructure could be placed so as to get the most bang for the buck in connecting the digitally divided to the global labor market. This story affected the measurement of stratified internet access and explanations for it.

I do not dispute the reality of these inequalities; very real gaps in internet access existed across class, race, and geography. But the narrow framing of these gaps led to a dominant understanding of *access* as the opportunity to compete in the New Economy. The access doctrine operationalized the problem of inclusion in the New Economy as two sides of a digital divide, one competitive, one not. Solutions to the problem flowed logically from there.

For the Clinton administration, economic growth was a question of making adequate investments into human capital: the skills and abilities making up what Gore called our "core capacity as human beings," those means of production internal to the laborer. This was often explicit. The 1994 Economic Report of the President made human capital investment the second administrative priority after deficit reduction. The report went on to clarify that "each American must be responsible for his or her own education and training" but that the state would make limited investments in education, training, and the internet infrastructure (e.g., fiber optics). These would connect Americans to training opportunities because "American workers must build the additional human capital they need as a bridgehead to higher wages and living standards" (Clinton and the Council on Economic Advisors 1994, 41).

At other times, this approach was implicit: the language of reskilling for knowledge work or connecting to online resources. Gaps in access were not crises just because PCs and internet infrastructure are necessary fixed capital for the New Economy, but because these technologies permitted access to reskilling opportunities that increased individuals' human capital, access to new markets for the products of human capital, and access to new markets *for* human capital. They made you competitive and allowed you to compete.

Human capital theory became a key concept for governance in the 1960s, as Adam Smith's theory of the term was reassessed by a new generation of economists and as planners sought to incorporate domestic educational costs and international development projects into the neoclassical investment theories that drove macroeconomic policy (Adamson 2009). Defining human capital as productive skills and abilities fixed to a person requires a mapping of its distribution, and the effects of investment in it, across increasingly larger scales and more fine-grained variables. Human capital theorists such as Gary Becker and Jacob Mincer provided the techniques to incorporate poverty management into the Keynesian political consensus. When that consensus shifted, so did the political use of human capital.

During the War on Poverty, the welfare state identified wage labor with independence, largely irrespective of the content of that labor. And so the state of dependency was feminized and racialized by its identification with paupers, people of color on the margins of the labor market, and unwaged housewives (Fraser and Gordon 1994). Independents received state support through payroll taxes (e.g., Social Security), while dependents received it through general taxation appropriated legislatively (e.g., AFDC). The latter were easily scapegoated because they were unproductive (i.e., they had low human capital). The neoliberal turn dismantled many of these programs, attacking both the recipients and the institutions serving them as destructive of human capital—not only disincentivizing people from pursuing waged work, but subtly destroying their ability to do any work at all. Where the earlier system gave starkly different treatment to "dependents" and waged workers, the worthy and unworthy poor, the emergent neoliberal state extended this human capital calculus to every potential participant in the labor market, not just the traditionally dependent.

While the right wing of the neoliberal assault framed these attacks in largely negative terms, destroying the fetters that held people back from their promise and punishing those who could not follow the new rules of the game, the left wing proposed a hopeful, creative solution. Digital tools and skills would offer new promise to those left out of the global labor market, increasing not just local or individual but also national productivity. What seems at first to be a technology policy is then part of a wider turn in poverty policy, wherein poverty-relief measures, like all public goods, became "neither pure commodities, nor pure public goods but new intermediate strains that combine features of both" (Fraser 1993, 17).

With subsidized PCs, modems, internet connections, and a dedicated measurement program to find where to put them, the access doctrine attempted to upgrade the nation's human capital stock so that the traditionally dependent, along with everyone else left behind by the economic dislocations of the 1970s and 1980s, could move out of low-wage service sectors with low productivity growth and into higher-wage knowledge work sectors that were, in the 1980s and 1990s, seeing big productivity gains. This is what would promote independence at the individual level and competitiveness at the national level.

The Clinton administration was thus willing to countenance limited state intervention into the "natural" functioning of human capital markets because of a post–Cold War spending pivot and a burgeoning alliance with Silicon Valley. This freed the administration to acknowledge that the market was not joining fixed capital computing resources to the human capital in need of upgrading quickly enough to transition US workers to the New Economy—all this before the term *digital divide* entered popular usage in 1996.

Although there is no consensus as to who coined the term, former White House staffer and MCI General Counsel Allen Hammond IV and Sesame Street Workshop cofounder Lloyd Morrisett probably used *digital divide* in the seven years between the passage of the High Performance Computing Act and the NTIA's 1998 *Falling through the Net* report (Eubanks 2007). It appeared nowhere in the 1995 edition of that report. Clinton and Gore used it while campaigning in 1996, comparing their investment in America's future to Dole and Jack Kemp's planned neglect of the same. *Digital divide* appeared four times, in quotations, in the 1998 *Falling through the Net* report and more than fifty times in the 1999 sequel.

During Clinton's presidency, the NTIA, a small wing of the Department of Commerce, released four increasingly larger, more fine-grained reports on the state of the digital divide in the United States in 1995, 1998, 1999, and 2000. At Gore's request, the agency had asked for the Census Bureau's monthly Current Population Survey to be updated to include household data on computer ownership and internet and telephone subscriptions. Results were then cross-tabulated by income, race, age, educational attainment, and region. The NTIA, and its reports picked up by Clinton and Gore on the campaign trail, became a key institutional ingredient in the emergent access doctrine by treating stratified access as a chief symptom of (and universal access as a logical solution for) the persistent poverty that haunted the overall optimism of the New Economy. It was here that the problem of human capital deficiency was operationalized.

The NTIA framed increased economic fitness as the goal of access and market competition as the means to extend access. The 1995 report found that poor, rural minorities were least likely to have a PC or modem, followed by poor Black residents of central cities—but that those positions were reversed when education was held constant. It decried this because those "most disadvantaged in terms of absolute computer and modem penetration are the most enthusiastic users of on-line services that facilitate economic uplift and empowerment."

Gaps in connection rates between White and Black or Hispanic households, even with income held constant, grew from report to report, with the 1999 report calling the digital divide a "racial ravine." This is another variation on the gap or canyon imagery of the early digital divide literature: a fissure borne of the New Economy, separating the "information disadvantaged" from opportunity on the other side (NTIA 1995). Gore often asked audiences to consider opportunities for access not just in rich suburbs but in nearby, poor, predominantly Black inner-city areas: Bethesda and Anacostia, Brentwood and Watts (Gore 1994a).

Each report ended by profiling the "least connected" who "lag further behind" and what they stood to gain through PCs and modems (1998). The 1999 report concluded, "While these items may not be necessary for survival, arguably in today's emerging digital economy they are necessary for success" (77) and "no one should be left behind as our nation advances into the 21st Century, where having access to computers and the internet may be key to becoming a successful member of society" (80). Policy proposals were absent in the first report but included in subsequent ones. Over time, they gave greater weight to market diffusion of the means of access, but they argued that time was of the essence and that "community access centers" such as schools and libraries could act as temporary bridges for disconnected communities.

A focus on the number of internet-connected PCs available dominated the access doctrine initially, but later coexisted with investigations of usage and skill, all broadly grouped under *access* (Epstein, Nisbet, and

Gillespie 2011). Access did not mean skills or tools specifically, but the general opportunity to compete. Bringing the digitally divided online was an urgent problem, not for reasons of human rights, religious obligation, or any of a variety of other possible frames, but because the root causes of crises of GDP were found in individual users and their PCs.

This research program quantifies the hope that the internet and personal computing would overcome poverty. And the *Falling through the Net* reports kickstarted a tremendous amount of digital divide research beyond the US federal government, led by researchers like myself who wanted to understand how new technologies were helping or hurting the poor. Questions of technological diffusion (e.g., Rogers 2010) were thus inseparable from questions of social mobility. The field owed much of this political urgency to the work of modernization theory. This literature largely conceives of progress in a linear, economically and technologically deterministic fashion, with governments and philanthropies engaged in a project with the goal "to bring the backwards people *forward*" (Graham 2008, 779; emphasis in original; see also Escobar 2012; Ferguson 1994). Early digital divide researchers largely supplemented the NTIA's work, measuring stratification of internet access and the success or failure of different deployments (Warschauer 2002)

In the 2000s, research began to address inequalities in not only which technologies of what quality were available to whom, but also the uses to which those technologies were put and the rewards drawn from them (Hargittai and Hinnant 2008). The focus shifted to the study of digital inequalities among those with access to the internet, with new research on equipment, autonomy, skill, social support, and the purposes for which the technology was employed (DiMaggio et al 2004). Richer accounts emerged of the mechanisms driving stratification: accounts, for example, of how managers force employees or those seeking work to develop particular skills (van Dijk 2005) or of the institutional and cultural dynamics that position Black and Latino youth as both marginal to the information economy and, through their informal learning practices, leading innovators within it (Watkins 2018).

Conceptually, social scientists integrated mid-level theories of digital inequality within broader accounts of stratification by geography, race, class, and so on (Robinson et al. 2015) by, for example, demonstrating that how young people use the internet, even holding basic access constant, varies widely by class. Higher-status individuals engage in more "capital-enhancing activities" online than lower-status individuals and thus reproduce their class position (Zillien and Hargittai 2009).

This progress in the digital divide literature is a recognition that social life does not conform to simple binaries between information haves and have-nots and that no technology on its own dissolves inequalities. Indeed, in our papers and conferences we found ourselves responding over and over to the simple binaries of the Clinton administration's measurement program. But that program's power went beyond its empirical findings. The framework of the human capital crisis exerted an inescapable political gravity, recruiting researchers to repeatedly refute, complicate, or nuance it—but never vanquish it. Even if it was poorly framed, the problem was too urgent to dismiss. The access doctrine compels scholars to respond. Its gravity draws us in.

But in the Clinton administration, this political urgency did not translate to large-scale political solutions. Digital divide policy was both technology and poverty policy, and neoliberal poverty policy appropriated new funds largely for police and prisons, not job creation or increased cash assistance. The NTIA's measurement program had to justify itself on this terrain. Its first report claimed that "once superior profiles of telephone, computer, and on-line users are developed, then carefully targeted support programs can be implemented that will assure with high probability that those who need assistance in connecting to the NII will be able to do so." The crisis of competitiveness was expansive, but the needs of the human capital deficient needed precise measurement.

Aid needed to be precisely targeted so that access would offer opportunities to compete and not handouts. If a relatively small number of Pell Grants for prisoners had been labeled handouts and canceled in 1994, then surely proposals to treat the internet as a public utility able to reach everyone in the country would be verboten later in the decade. Solutions would instead come from the deregulation of telecommunications markets and carefully targeted interventions within community access centers.

The Digital Divide and Its Market Solutions

The Clinton administration's discussion of access solutions became a meditation on state limits. This final part of the access doctrine made distribution of these solutions the responsibility of deregulated markets, wherein competition would lower prices and extend access. This forced a reconsideration of the universal service mission—the provision of baseline connectivity to every citizen in the name of safety and political and economic participation—in telecommunications policy. Consistent with other contemporary neoliberal projects, state intervention would persist, but only insofar as creating markets and securing competition in them. In the meantime, community access centers would triage technological poverty. These solutions cement a master definition of *access*: the opportunity to compete in the New Economy, an opportunity independent of, but able to strategically mobilize, any individual digital technology. It is this master definition that continues to animate the politics of and research into access, even as the specific questions addressed have expanded far beyond who has in-home internet. Although the access doctrine resolves the contradictions of neoliberal poverty policy, any possible solutions that emerge from it are necessarily constrained by the push for private sector promise and public sector punishment.

The administration used technology policy to stake out the purpose and limits of the state during economic transitions. Press releases for the Next Generation Internet Initiative even included Q&A sections asking why the government was involved at all (Clinton and Gore 1996a). This was posed as a reaction to a larger economic problem beyond the government's control. The NII *Agenda for Action* (Department of Commerce 1993) described a new era in which "information is one of the nation's most critical economic resources" in every industry trying to thrive "in an era of global markets and global competition." Its future priorities are listed under the heading "Need for Government Action to Complement Private Sector Leadership": tax and regulatory policies that promote long-term private investment, universal service, and research programs and grants that help the private sector build and demonstrate NII applications.

Laissez-faire economics is never a hands-off approach, but always an activist policy wherein the state is charged with creating and protecting markets. Plans for a Global Information Infrastructure (GII) that would end the global digital divide hinged on the request of the World Treaty Organization (WTO) for member states to privatize state-owned telecommunications (Clinton and Gore 1995). Gore (1994b) compared the GII's promise to the contemporary privatization of the former Soviet Union's telecommunications infrastructure, arguing that "reducing regulatory barriers and promoting private sector involvement" allowed freedom of movement for information, capital, and democracy.

Prioritizing market creation would seem to contradict the universal service mission that the 1995 *Falling through the Net* report argued was "at the core" of US telecommunications policy. After all, there is always someone who cannot pay, always an area where new infrastructure is too costly. This universal service mission emerged from early twentieth-century competition between the first US telephone companies, which refused to connect to each other's customers. The 1921 Willis-Graham Act admitted that "there is nothing to be gained by local competition in the telephone industry" and permitted AT&T to form a monopoly that eventually spanned the country in exchange for a commitment to cover as much of the country as they could. The 1934 Telecommunications Act created the FCC to regulate telegraph, radio, and telephone traffic and to negotiate with AT&T over price controls and service quality (Kim 1998). State-enforced private monopoly guaranteed universal service—exactly the sort of anticompetitive, Keynesian compromise the Clinton administration argued was disrupted by information technology.

This conflict was resolved by selecting certain aspects of the universal service mission, particularly its identification of individual ownership of technology with democratic participation and economic security, for incorporation into a broader story about market creation and participation. This meant equitable access would be best facilitated not by monopoly, but by cross-media competition. In this way, the regulatory apparatus that in an earlier era ensured universal service became an enemy of that same mission.

By the 1999 NTIA report, universal service was largely a stopgap measure for "high-cost areas" left out after a program of "expanding competition in rural areas and central cities" (78). Here, in the last report with *divide* in the title, universal service was a question to be asked after pro-competition policies were realized. This was foreshadowed by a 1994 Congressional Research Service report showing that Gore's original nine principles meant to guide NII policy were, a year later, cut to five. Gone was the explicit universal service principle, replaced with a new commitment to not creating "information haves and have-nots."

This commitment registered not as universal service but as an emphasis, increasing over time, on triaging the digital divide through community access centers such as schools and libraries. In 1995, such centers were temporary "safety nets" in a "long-term strategy" (NTIA 1995, 6). But by 2000—and despite a report ten times the first's length, which stressed that "not having access to these tools is likely to put an individual at a competitive disadvantage" (NTIA 2000, 89)—the NTIA observed the increased use of libraries by the un- or underemployed without any judgment or policy proposal. It was a settled state of affairs. The later reports had faith not only in the competitive boost information technology provided the poor but also in the power of markets to extend those opportunities. Indeed, the 1999 report reinterpreted history to fit this script, comparing internet and telephone buildout and arguing that "high levels of telephone connectivity" were achieved primarily through "pro-competition policies at the state and national levels" supplemented by universal service subsidies—rather than the monopoly granted AT&T (NTIA 1999, 77).

Universal service was always more of a political principle than a specific set of proposals and objectives, vulnerable to reframing. Crawford (2013) describes the 1990s reorganization of US telecommunications as an anticipation of the possibilities of media convergence and a reaction to monopolies borne of Reagan-era rates deregulation. Trying to manage burgeoning oligopolies, the 1996 reform of the 1934 Telecommunications Act pursued universal service largely through further deregulation. Cross-media competition and ownership was permitted in all markets; local phone companies could offer long distance, cable companies could offer internet, the Baby Bells borne of AT&T's break-up had to let smaller companies offer services on their circuits, and all cable rate regulations were ended.

Internet access for schools and libraries would be supported by the Universal Service Fund, administered by the FCC from taxes, and the E-Rate subsidy, collected from telecommunications firms—an easy target for court challenges (Hammond 1997). There was no similar provision for households. Indeed, the FCC later argued that compelling firms to offer services of equal quality or speed in rural and urban areas "would undercut local competition and reduce consumer choice and, thus, would undermine one of Congress's overriding goals in adopting the 1996 Act" and that equality should therefore not be considered as part of the universal service rubric (FCC 1997).

At its core, the creation and protection of markets as a neoliberal political strategy relies on the idea that more competition will bring more winners and fewer losers (Dean 2008). Clinton could promote free-trade agreements while warning about the need for workers threatened by globalization to push for reskilling because both were framed as competitive responses to New

Economy stakes. This competition for competition was how the Atari Democrats revised the party's postwar social democratic agenda while embracing the punitive turn of the right. It was in this context that the digital divide was named and conceptualized, and we have yet to escape that legacy. Even as scholarship and policy matured and began to concern itself less with access to goods and more with skills or rewards, this original framing of the problem ensured that no matter how *access* was operationalized, it still denoted an opportunity to compete in the global economy, best provided by competition to offer that opportunity.

Conclusion: The Future of Access

The Clinton administration ensured that discussions of digital equity were understood as a problem of extending digital lifelines to the information economy: an opportunity to compete in the future, rather than being left behind in the past. But the solutions to the problem, particularly in the 1996 Telecommunications Act, that flowed from this framing ensured that it would not be solved, even on their own narrowly defined terms. A deregulatory approach resulted in a highly concentrated market, to the point where 85 percent of US consumers today have zero or one choice for high-speed internet offering 100 Mbps downloads, and 43 percent have zero or one choice for 25 Mbps downloads (FCC 2016).[4] The most recent data show that of the thirty-seven Organization for Economic Cooperation and Development (OECD) countries, the United States is sixteenth in home broadband subscriptions per one hundred people—below not just the Nordic social democracies, but also comparatively poorer nations like Portugal, Latvia, Estonia, and Greece (OECD 2019).[5]

Nor has increased competition brought better or cheaper services. The OECD divides mobile subscriptions into low-, medium-, and high-use plans, and at each stage, US subscribers pay almost twice as much as the OECD average. The story is roughly the same for in-home broadband, with low-end US subscribers paying around forty-six dollars per month compared to an average of about twenty-eight dollars, and high-end subscribers pay about sixty-one dollars per month versus an OECD average of about thirty-four dollars. And the birthplace of the internet has pretty slow internet. Although the United States is eighth in the OECD with an average speed of 18.7 Mbps—still well below South Korea, Japan, and the Nordic countries—that ranking falls to sixteenth when examining the different speed tiers into which fixed broadband subscribers fall. Only 4.1 per 100 Americans have 100 Mbps connections, compared to 18.5 in Switzerland and 11 in Belgium, and 7.2 per 100 Americans make do with 2 Mbps internet, compared to 3 and 3.8 in Switzerland and Belgium. It makes sense. What incentive does Comcast or AT&T have to upgrade its broadband network when so many consumers don't have a choice in the first place?

Because the access doctrine joined poverty policy and technology policy, its effects were felt not only in the regulation (or lack thereof) of private, consumer internet but in the plans (or lack thereof) for public alternatives. State legislators have repeatedly passed laws forbidding localities from building low-cost municipal broadband operated as a utility (Koebler 2015). In 2018, an FCC commissioner labeled these public resources a threat to free speech (Brodkin 2018). Although this would seem to run counter to the goal of increasing competition, it's important to remember that Clinton and Gore's story positioned the state only as a guarantor of market competition, never a participant. This dynamic was in full view at the Connect.DC event that opened this chapter. The city used Recovery Act funds to build a high-speed municipal broadband network to serve social service agencies in poor, majority-Black areas that larger internet service providers neglected, but it was never an option to extend those connections to consumers.

As we've seen, community access centers like schools and libraries were viewed as stopgap measures serving the fringes of the market before superior, private, home-based alternatives arrived. This led to a short-term approach to planning for and funding these community spaces, defined by intermittent cycles of competitive grant applications (Viseu et al. 2006) and lower levels of funding compared to other developed economies (Jayakar and Park 2012).

On the ground, this meant that the United States largely avoided the telecenter phenomenon embraced by other countries—the public equivalent of internet cafes, also missing here—while piling more obligations onto the schools and libraries that already act as community centers. Eighty-nine percent of DC public libraries, the focus of chapter 3, reported in 2012 (the last year for which data is available and the first year I began visiting MLK) that they were the only source of free internet in their area—a level higher than any state (Bertot et al. 2012). In a 2014 national survey, 77 percent of Americans who lacked internet access at home said that access at the library was very important to them or their family. Pew characterized this as part of a shift in library identity, from "houses of knowledge" to "houses of access" (Pew Research Center 2014).

Data on user behavior backs this up. Kinney (2010) found that the presence of internet-connected terminals (as opposed to their absence) has a significant positive effect on a library's total visits and reference transactions. Because both public schools and libraries are largely funded by state and local taxes (the latter significantly more important for libraries), this increased service load is not met by increased funds in times of crisis; as recessions hit, more people require public services, tax receipts dry up, and the funds available to the institutions decrease. Internet access and digital skills training are supposed to provide a lifeline in moments of economic uncertainty, but it is precisely during these moments of high need that community access centers have less to work with.

It didn't have to be this way. That other countries avoided this particular neoliberal marriage of technology and poverty policy shows that the US approach was not inevitable. Straubhaar et al.'s (2008) comparison of US and Brazilian technology policy makes this clear. The authors found that the Clinton administration focused primarily on physical access and framed technological stratification primarily in terms of economic opportunities lost in an inevitable moment of economic transition. The contemporaneous Brazilian *inclusao social* framework made access one part of a broader social mission to overcome long-standing inequalities based in race and class.

Brazil's Cardoso government set the goals for access policy in its 1997 Green Book: new research initiatives in science and technology, distance learning, cultural preservation, telemedicine and the modernization of health systems, the construction of local e-commerce platforms, and technology education at all levels. The state was the primary actor in this frame, and the citizens in their community (rather than human capital in the market) formed the primary site of intervention. It was an explicitly political framework, increasingly so as Lula's left-wing developmentalist government took over in 2003.

This naturally led to interventions different from those pursued in the United States. Brazil's universal service fund collected 1 percent of telecommunications firms' revenue, rather than the variable contributions levied on US firms based on their own quarterly revenue projections. These funds were directed not only toward schools and libraries, but toward direct infrastructure investment, assistive technologies for the disabled, and the creation of purpose-built telecenters providing wraparound social services through partnerships with local civil society groups. Local municipalities funded telecenters and provided technical support, civil society groups managed them, and the whole process was administered by a community council of local telecenter users who ensured that the initiative catered to local needs. Telecenters ran on open-source systems to reduce licensing fees and maintain the spirit of democratic participation. National competitiveness was never entirely out of the picture but, because of the broader work of a developmental state emphasizing historical inequalities, it was subordinated to community control and community empowerment. Access was a social good, not an opportunity for competition. Such an idea could not fit within the US access doctrine.

For the Clinton administration, and subsequent reformers working under that script, new inequalities were borne of skill gaps in the New Economy, rather than long-term problems of deindustrialization exacerbated by punitive poverty politics. Indeed, the power of the access doctrine comes in large part from how these technological solutions reconcile the contradiction in US poverty policy between promise and punishment. A global communications network makes opportunity available everywhere. Those that do not choose to log on and take part become a drag on regional and national productivity and so must be punished.

In his final State of the Union address, Clinton (2000a) told the nation, "We have built a new economy." Brought into office during a recession and after the collapse of the USSR, his administration was supported by export-oriented technology industries prepared to countenance mild state economic intervention that would catalyze private investment in internet infrastructure and upgrade US human capital stocks for the New Economy. This economic nationalism would create and protect markets and ensure participation in them but lacked the direct job creation or public works of prior Keynesian regimes. The access doctrine managed the anxiety of flexible economic relations by positioning access not just as a tool or a skill, but as the opportunity to compete in the global network. Even when the actual distributive mission of increased access narrowed over time, it continued to effectively frame the problems of poverty in the New Economy not as dislocation borne of deindustrialization or the retreat of the welfare state, but as the absence of investment—by state or citizen—in human capital and the technologies to grow and market it.

But the problem with an approach to equity based on sound investments in human capital is that a new set of investors can just as easily declare them unsound, which is what happened when George W. Bush entered office. His FCC Commissioner, Michael Powell, famously riffed on the persistence of the digital divide: "I think there is a Mercedes divide.... I'd like to have one; I can't afford one" (Labaton 2001). This signaled a shift that included prominent cuts to an Education Department program funding community access centers and a Commerce Department program for underfunded organizations, like food banks, attempting to modernize their infrastructure (Schwartz 2002).

In response, representatives of liberal think tanks like the Benton Foundation argued that this political retreat kept the nation from leveraging sunken investments that could effectively mobilize human capital (e.g., Wilhelm 2003). But the "Mercedes divide" comment was not fundamentally at odds with the script set by the Clinton administration. Powell just held that this sort of capital investment was unnecessary to increase individual or national competitiveness. Prior investments had matured. Further funds would be wasted. The script persisted, even as the left edge of neoliberalism weakened and the right strengthened: equity was still a problem of human capital investment; it was just no longer a good investment.

Hope, then, crosses the political aisle. What differs between the right and left wings of neoliberalism is the particular emphasis placed on the carrot or the stick in the development calculus. Powell's dismissal of digital divide politics was not a dismissal of the crisis itself or even the terms of the crisis as set by the Clinton administration; after all, the Mercedes divide comment came alongside further deregulation of telecommunications markets. Powell agreed that "it's an important social issue" but suggested that the diffusion of devices specifically would largely be taken care of by the free market (Lasar 2011).

This move was consistent with broader changes in federal poverty policy under the second Bush presidency (Allard 2007). The neoliberal shift from cash assistance to workfare programs continued apace. Cultural reengineering of poor people's human capital intensified, with new marriage promotion programs and a prioritization of faith-based organizations in poverty relief. Block-granting of poverty relief funds to states was reauthorized. This made the recovery from the 2008 recession even more painful. An emphasis on increasing competitiveness through punitive poverty policy thus continued, even as the digital divide program was largely offloaded to the private telecommunications market. This should be seen not as a failure of the access doctrine but as a success. Targeted investments were no longer necessary because opportunities for competition abounded; the state needed only to secure the grounds for competition.

The production and reproduction of the access doctrine does not, of course, just happen from the top down. This hopeful story had its origins in the Clinton administration, but the problem of poverty only becomes a problem of technology when the institutions managing technology and poverty teach themselves to make that link and pass the lesson on to the rest of us. The remainder of this book explores this process, away from the halls of power in Washington, DC, and into its streets, classrooms, and office buildings. The access doctrine set new terms for social reproduction. The institutions that teach us how to make a living changed in response, and in so doing they taught us that new technologies and new skills will secure our economic futures.

Cities were particular sites of concern for neoliberal poverty policy and economic development. In this new regime, human capital stock is not just imported from outside but upgraded within, in the likeness of gentrifying outsiders. This "creative city" (Florida 2004) is never separate from its punitive shadow. As Spence (2015) argues, urban entrepreneurialism delegates risk-taking and responsibility to cities and individuals at the precise moment that cities' tax bases shrink and they are forced to compete with one another. Safety nets grow more restrictive and carceral solutions become more common, with one effect being the radical restriction of acceptable entrepreneurialism (e.g., heavy punishments for drug sales or unlicensed food vendors).

Municipal leaders may wish to do something different when it comes to economic development, poverty relief, or a host of other fields, but the options available to them are limited both ideologically—there is a limited menu in their broader political networks—and materially—limited federal and state support exists for solutions that don't involve luxury housing, tax breaks for relocating corporations, and heavy-handed policing (Forman 2017). The hopeful but limited vocabulary at the OCTO event that opened this chapter is one sign of these constraints.

The problem of persistent poverty in the information economy, both for individuals and regions, is overwhelming, and the influence of the access doctrine provides a way to understand it. The reduced resources available force urban institutions to quickly reorient themselves around new technologies and new goals—to bootstrap. In doing so, they model themselves on the ideal-type organization of the New Economy: the internet startup. Just as individual economic actors are trained to remake themselves as technological entrepreneurs in order to succeed, so too are training organizations. Chapters 3 and 4 examine this process in schools and libraries, exploring how their quest to close the digital divide changes who they are and what they do. But to really understand this process, we need to start with the "right" side of the divide and the sort of hopeful organizations and people everybody else is trying to become: tech entrepreneurs and their startups. It is to them, and their place in the city, that we now turn.

2   The Pivot and the Trouble with "Tech"

I first met Travis and the rest of the InCrowd team at their office party in February 2014. They were celebrating their move out of rented incubator space and into a space of their own. Framed software trademarks met guests as they entered through the glass double doors, just in front of the conference room where a huge company logo hung over a table of hors d'oeuvres. Most employees wore orange and black InCrowd tees, with streamers in the same colors flying above. Beer and cocktails flowed from an impromptu bar set up in the back next to the DJ booth, and a magician roamed the halls doing card tricks. Founder Travis gave a tour to anyone who asked. In his usual blunt fashion, he noted that this was as much a recruitment event as anything else, showing that InCrowd had arrived and was ready to act on his "number one operating principle": "being an employer of choice for the right people." A developer led me toward the Wall of Awesome Ideas, which recorded brainstorms for outlandish products. It was filling up fast. I shared a toast with a new hire, who said she was celebrating her own transition: from the slow, predictable pace of nonprofit work to a tech startup, where every day brought a new challenge and demanded new skills.

It was a night of hopeful transitions: new space, new careers, and a new identity for the startup. The move coincided with InCrowd's pivot from a business-to-consumer (B2C) company, with software that supported anyone planning a party, to a business-to-business (B2B) company building software that specifically supported caterers in planning how food, drink, and place settings for large events were bought, scheduled, prepared, presented, and stored. It was a tough transition. Beyond fundamental changes to the software, all of the company's marketing and training materials had to be substantially revised, but everyone was still excited. The company was supported in the change by a $2 million seed-funding round from the previous summer, an amount that would be quadrupled with its Series A investment at the end of the year.[1]

For Travis, the welcome party was about supporting this shift and changing the public perception of his startup. More than anyone else, he internalized the change. As the CEO and cofounder, he felt he needed to campaign for this big change, especially for potential investors, customers, and employees: "People buy me. Especially when they choose to work here, they want to believe in who I am."

In startup land, this is called a *pivot*. It's a gut-check moment for a company, where survival depends on fundamentally reorienting how you do business, with whom, and why. This idea is everywhere in tech, but its most powerful formulation comes from Eric Ries's *The Lean Startup* (2011). Inspired by the lean production model pioneered by Toyota in the 1970s, Ries says startups should be constantly iterating new prototypes, pushing them out the door as soon as they are ready. For Ries, a startup doesn't necessarily produce software. It's bigger than that, "a human institution designed to create a new product or service under conditions of extreme uncertainty" (27). Extreme uncertainty and how to manage it—and the thrill of doing so— is a theme throughout. Unlike traditional businesses, startups don't necessarily know who their customers are, what their best products are, or how to bring the two together. This environment where the work process changes constantly based on the data is not only accepted but embraced.

This is life on the "right" side of Clinton and Gore's digital divide. In startups, the access doctrine appears to come true: the right tools and skills give you a chance at weathering this environment of radical uncertainty. With the right data, and the right mindset to make use of it, organizations can not only survive the economic storm winds but master them, using the momentum to reach new and greater heights of opportunity.

The access doctrine tells us that those on the margins of contemporary capitalism will move to its center if they have the right digital tools and skills. As we saw in the previous chapter, this political common sense emerged out of the Clinton administration's attempts to manage the persistent poverty of the information economy. The access doctrine is a hopeful story that justifies both the reduction of the welfare state (because it held human capital back from full utilization) and the expansion of the carceral state (because threats to regional or national productivity must be contained). The world is divided into digital haves and have-nots, but the opportunity available through the internet promises to close that gap.

This idea holds special power in moments of economic instability, and so the high-tech people and organizations who thrive in these moments become models for everyone else. (Not coincidentally, they are largely White professionals who are also an important part of the new Democratic political coalition.) For individuals, the access doctrine presents a clear path out of poverty: learn to code like InCrowd employees and become an entrepreneur like Travis. For organizations that address the problem of poverty and embrace the access doctrine, InCrowd's flexibility becomes a script for success. I use *bootstrapping* to describe schools' and libraries' attempts to follow this script. Bootstrapping is the process of rapid organizational restructuring that occurs when public service organizations define the problem of poverty as a problem of technology and the skills to use it. Bootstrapping begins when public service organizations are faced with overwhelming problems, limited resources, and diminished legitimacy. They rapidly change their identity and operations to provide the people they serve with digital opportunities. This process is modeled on startups' pivots, but schools and libraries are simply different types of organizations, ones that cannot pivot like startups.

Beyond individual organizations, startups also offer hope to cities scared of falling behind in the global economy. Because of their cultural dynamism and their employees' disposable income, startups appear to be a route to regional economic security. And so cities compete to remake themselves in tech's image, in an effort to grow tech at home and attract it from elsewhere. At each scale—individual, organizational, and structural—the startup and the entrepreneurial citizens (Irani 2019) it employs become models of success in moments of extreme economic uncertainty. This chapter describes the internal functions of that model, showing how central the pivot is to the cultural life of startups—for individuals, firms, and the sector as a whole—so that later chapters can explore attempts to generalize it.

The moment of the pivot is where all this uncertainty comes to a head. It is, in Ries's (2011) words, "a structured course correction designed to test a new fundamental hypothesis about the product, strategy and engine of growth" (149). InCrowd was undergoing a "customer segment pivot." Travis and his team had realized that their product best served enterprise customers—professional caterers and event planners—rather than regular consumers planning the occasional graduation party or quinceañera. But. the pivot meant much more than a change in who the company sold software to. It also meant a fundamental reconsideration of what the company was and where and how it worked.

In the name of progress, internet entrepreneurs like Travis—indeed, all of InCrowd—must be prepared to discard their old tools and ideas, or at least radically restructure them based on the data. Ries even goes so far as to redefine a company's survival prospects—its *runway*, usually defined as the amount of cash on hand divided by the rate of spending—as the number of pivots it has left: its remaining chances to learn, rebuild, and radically reinvent. The pivot is not just a corporate strategy, it is an identity. This capacity for reinvention is important for the internal and external legitimacy of these firms, generating buy-in from the employees that put in long hours and the cities that put in tax breaks. Ahead, I explore this legitimation process at the level of founders, firms, and cities. Startups today serve as an "ideal type" for public service organizations under pressure. But, of course, ideal types do not correspond to empirical reality; they are abstractions built for explanatory purposes. This abstraction—focused on the ability to pivot—serves different purposes at different scales of the tech sector.

This chapter builds on pioneering ethnographies of the startup scene during the turn-of-the-millennium dot-com boom, particularly Gina Neff's (2012) exploration of venture labor. For Neff, *venture labor* is a strategy that professionals follow for managing the risks inherent to the contemporary labor market. They take an entrepreneurial approach to their own careers, viewing their work in a particular company that may not be long for this world as an investment in their creative portfolio, human capital, or financial security. In many ways, this chapter documents the normalization of what Neff identified as an emergent strategy in 1990s and early 2000s Silicon Alley, New York (see also Marwick 2013; Ross 2004). These are now accepted ways of living and working, with their own ecosystems of cultural and educational organizations to train new entrants to the sector.

Founders' Pivots

Founders internalize the identity of their startups in order to demonstrate the vitality of the firm to powerful outsiders and uncertain insiders. They learn to do this from their peers, from financial pressures on the firm, and from the bridging organizations that tie local, national, and transnational tech communities together. Travis was always up front in saying, "the company is a reflection of me." This was not self-aggrandizement. Angel investors conduct their due diligence on a founder's background and management style as much as the fundamentals of an early-stage company with great promise but little to show for it. Founders also act as the firm's ambassadors to politicians, clients, and partners. Identifying with one's firm and performing that for outsiders necessarily means that pivots are deeply felt—as questions not just of business strategy but identity. Travis was not the only founder I met that went through this process.

I kept up with entrepreneur Ji for several years as she built Hearth, a matchmaking app for conversation partners. She knew monetization was inevitable—you have to pay the bills—but she worried that generating coupons to draw conversation partners to a bar or restaurant, and all the sales work that would require, would necessarily distract her from solving our "problematic decline in connectivity." Ji was just as invested in researching connections she saw between "gun violence or digital divide issues or political chaos" as she was in bringing Hearth to market. Monetization was a necessary pivot, something that sullied Ji's mission even as she knew it was necessary for her app's long-term health.

Ji's other pivot was geographic. Hearth had grown in DC, piloted with a series of group dialogue sessions hosted by local high-end restaurants in rapidly gentrifying areas like Shaw, curated around themes like trust. There, Ji tested out some of the prompts Hearth would use for in-person conversations and the curated group of mostly White professionals who repeatedly voiced a desire for more unplanned social connection in lives marked by eighty-hour work weeks, constant internet use, and frequent travel. It remains a cliché that White professionals in DC are rarely from DC, and that made the city the perfect venue for Hearth. The hopeful, cultural mission of the startup was thus perfectly aligned with Ji and her social network, who said, "For the first time in my life I really felt like I was among my people here in DC." But she knew the investment networks for consumer technology were stronger in San Francisco, and a move was on the horizon. It hurt because she was leaving her "tribe," the cultural geography that had birthed Hearth.

Founders like Ji and Travis do not learn to internalize their firm in a vacuum. They are trained to think of themselves as equivalent to their company by the bridging organizations[2] of the startup ecosystem. These organizations are not startups themselves but build a sector for them. In DC, the 1776 startup incubator in Logan Circle is one of the most important sites of sector-building, as well as something of a meeting point for various smaller communities within the scene. Evan Burfield and Donna Harris founded 1776 in 2013. Their office space was secured by Mayor Vincent Gray, who often framed 1776 as a cornerstone of his five-year economic development plan. David Zipper, who became a key economic development advisor for Gray after holding a similar role for Michael Bloomberg in New York, led the creation of the DC Tech Incentives program, along with 1776's dedicated $380,000 grant from the city. Later, 1776 would hire Zipper to lead its cities and transportation portfolio. In this role, he advised startups focused on urban problems, and lobbied municipal governments on 1776's behalf.

Burfield made his name as an executive in consulting firms. Harris began her career as a serial entrepreneur in Detroit. She said that she "knew then that startups like mine could have been critical job creators to a city in decline. But I didn't have data to prove it, and I didn't have a platform to convince anyone I was right or strategies to bring about change" (Harris 2016). After working in enterprise software sales, she was appointed managing director of the Obama White House's Startup America initiative, a public-private partnership to fund, train, and lobby on behalf of startups and cities recruiting them. The 1776 incubator rents office space to startups lacking it, provides training and resources to them, and hosts parties and informational events for the larger startup community.

There were other training venues that established the rules of the game besides 1776. The various WeWork coworking spaces were important here. One way that WeWork justifies its rental costs—$400 to $550 per month for a dedicated bullpen desk—is through the events (e.g., "How to Properly Offer Unpaid Internships") and networking opportunities available to tenants. Tech Cocktail, run by former AOL employees, ran an industry news website and sponsored events, like pitch competitions in the style of the *Shark Tank* reality TV show. Various bootcamps like General Assembly promised to quickly reskill people for the information economy, taking disgruntled bureaucrats and turning them into in-demand developers. Mentoring also occurred in informal meetups, where founders or developers with particular skill sets got together once a month. Or it occurred through designated advisors who brought their years of experience to younger founders—sometimes in exchange for equity, sometimes not. And though

DC tech workers and executives often bemoaned the lack of a large technical university like Stanford or MIT around which the scene could orbit, local universities (e.g., George Washington, American, Maryland) still brought entrepreneurs together for hackathons, networking, recruitment, and more.

Tech Cocktail cofounder Frank Gruber released a book in 2014 entitled *Startup Mixology: Tech Cocktail's Guide to Building, Growing, & Celebrating Startup Success*, blending business advice with self-help for budding founders. Although it lacks the scientific pretensions of *Lean Startup*, it was similarly meant to help readers get into an entrepreneurial mindset. At one event held at Crystal City incubator Disruption Corp, Gruber pitched the book as a twenty-first-century successor to the self-help classic *How to Win Friends and Influence People*. He invited several founders on stage to discuss their failures and successes. Each was greeted with a cocktail personalized to them and their firm. Mixologists made the drink, passed it up to the founder on stage, and then passed samples throughout the crowd.

I was in the audience with Travis, who had waved at me from across the room as I entered, furiously typing emails on his laptop before the panel began, clad in his usual jeans, sportscoat, and InCrowd t-shirt. As people mingled after the event, he related InCrowd's origin story to a small group. Travis told them he'd been frustrated with overcrowded, understaffed parties in his native New York during the dot-com boom years, when he was working as a web developer. He omitted the portion of the story he had told me in an earlier interview, where he joked about first wanting to design an app that would make sure he always sat next to the prettiest girl at friends' weddings. That was a few pivots ago.

Travis was a rising star in the scene. He was one of the first graduates of The Fort, another incubator and predecessor to 1776. And 1776 had cut InCrowd a deal on rent before it moved to its current location, something Travis described it as a mutually beneficial relationship because InCrowd's fast growth lent the new incubator an air of respectability. InCrowd had been a side project while he worked on his MBA and then for a government contractor, whose impersonal bureaucracy frustrated him to no end.

This story was important to InCrowd. Travis's identification with the firm was not just a performance for outsiders, it was also a management practice. Everyone needed to identify with the company's pivots. Travis told me in an interview, "I'm a majority owner, everybody here is an owner [i.e., all InCrowd employees received a small amount of equity] so everybody here is an entrepreneur. I know it sounds cheesy, but everybody here is an entrepreneur. Everybody is entrepreneurial....My goal is to make it everybody's baby, but I might have the head. They might have a finger."

Travis needed to be a role model in an environment of extreme uncertainty, and he told mid-level managers he expected the same from them, because he wanted his employees to commit to their new business model, and all the various smaller challenges that would arise from it. The idea of the pivot was central to the InCrowd's identity and operations. Employees circulated the idea as part of their efforts to build a company culture, and in the process learned to embrace a great deal of flexibility not just in the organization's structure and purpose, but in their day-to-day or even minute-to-minute work habits.

The Ever-Pivoting Startup

When I asked Christopher, a bootcamp founder, what he was trying to teach his students, he didn't hesitate to say "entrepreneurialism." I asked what that meant. He replied: "I think to be an entrepreneur really encompasses an orientation towards being able to think on your feet, to constantly pivot, to work in constantly changing dynamic environments, to observe the world and the changes going on around you and come up with solutions and business ideas that take advantage of the opportunities in front of you—not necessarily shorter term, but very quickly growing business ideas." There is nothing in this definition about technology. "Business ideas," Christopher would make clear, could be a new approach to a design problem or a particular way of handling customers. This is a philosophy of work: surviving and thriving in an environment of extreme uncertainty by taking in new data and using that data to quickly change your work habits. The problem he was hoping to solve, to return to the access doctrine, was much bigger than a skills gap. It was a culture gap.

This is exactly what Travis hoped to instill in his employees. Indeed, he hired some of his first employees from these sorts of bootcamps precisely because they had the correct mix of skills and philosophy—and so didn't need a lot of on-the-job training. The company grew quickly, but there was every indication that Travis's cultural project worked, and not because they blindly accepted his instruction. InCrowd's culture was something everyone was self-consciously engaged in building. Organizational culture within startups is built on an embrace of large-scale strategic pivots, in the sense in which Ries described them, as well as a sort of daily pivot between different work tasks and the different mindsets that support them. Embracing flexibility legitimates an environment of economic uncertainty and the long hours and unclear prospects that entails. Who teaches techies to embrace flexibility? The previous section discussed sector-building organizations outside the firm. Within the firm, women encourage this embrace as part of the startup's general culture-building project. Such assignments either occur in the formal capacity of nontechnical roles, as one part of their regular duty to manage the emotions of peers and clients, or in the informal capacity of unwaged emotional labor women are expected to carry out in waged work because of the regular, unstated assumption that they are naturally good at it (Hochschild 2012; James 1989).

Beneath Travis and his assistant Karen, InCrowd was made up of three teams: Design developed the software; Sales secured new clients; and Contact trained clients in the use of InCrowd software, designed templates for them, answered and managed customer complaints and questions, collected outstanding bills, and sold existing customers on new products and bigger contracts. Beth led Contact. She had conducted her own personal pivot away from her dreams of being a speech pathologist and into a career in tech. InCrowd was her second startup. She told me, "I've been here for seven months, but on a day-to-day basis, my job has probably changed four or five times. It's really exciting." She would never dream of leaving tech now, largely because she had learned to embrace the environment of extreme uncertainty and the personal and professional pivots it required of her: "With startups, you do a lot of personal growth because you kind of are thrown into a role that's sink or swim. You're either going be able to do the path and then some and move with the pace of the company or you're not going to be able to handle it. It's pretty ambiguous."

Her previous workplace had, if anything, more of the stereotypical startup aesthetic than InCrowd's: games, frequent happy hours, a ball pit. But it didn't *feel* like a startup to her because of the emergent bureaucracy that segregated employees into clearly defined roles and kept them there all day—itself an organizational shift made in an effort to sustain their early growth. That failed. Beth was eventually laid off as nontechnical roles were outsourced.

At InCrowd, Beth's days were a series of pivots in the kind of work she did and the kind of relationships she had to maintain through that work.

For her, this was largely about how she dealt with clients over the phone through careful choice of tone, words, and diction. The default was a breezy, instructional demeanor as she walked party planners through a particular piece of their software. There were no bad questions. She was inviting people to explore the product with her in order to build it into their routine: "This is going to make your life easier, I'm more of a visual person myself....I'm very impressed, it looks like you have that tool down." But this could change fast. Sometimes she had to finish one teacherly call only to turn around and play debt collector. This required a sterner reprimand. Throughout, she was also careful to request more information on various events, menus, and staffing arrangements that might be useful for future pitches to the client.

Variations on these strategies were also deployed internally. Beth would advise subordinates on key phrases (e.g., "Please advise...") to deploy to get the desired effect and coach them on the strategy they would take for specific clients. Stiffer stuff was deployed to get InCrowd developers to follow up on a particular feature implementation or when graphic designers hadn't fixed something to a client's specification. She was a leader who liked leading, but the emotional pivots required of her were exhausting. It wasn't rare to see her arrive at her desk at 8:00 a.m. and not leave until 9:00 or 10:00 p.m.

Vijay, a lead developer in Design, was also drawn to tech because of his romance with the atmosphere of constant change. He had tried his own startups and failed and was happy to now be mentored by Travis. "Because we're all startup people," he said, "we all do a bit of everything, from survey system administration, managing CAD infrastructure, to how many pixels wide this particular element should be, the whole thing." Like many people in DC tech, he would favorably compare startup life with his prior experience in large corporate or federal bureaucracies. And in contrast to Christopher's lofty philosophy that took entrepreneurialism beyond coding, startup culture was inseparable from software development for Vijay. For him, mastering the environment of extreme uncertainty was best achieved through the immediate feedback you got from building and implementing code. This is why he left policy work: "I thought that the opportunity to actually close that feedback loop, to actually have an effect on my job, on the world that I was living in, was much, much tighter if I were doing [software] development."

Where Beth's hour-to-hour changes were often about evincing particular emotions from clients and subordinates, Vijay's were about corralling a quickly changing product. InCrowd sold catering software. New features were constantly added to support new needs, especially from big clients, or to solve unanticipated problems. Major redesigns had to be built with prospective clients in mind, and this required research and conversations with Sales about sectors they hadn't yet reached and what sort of silverware, plumbing, waitstaff, and so on they might have.

Over the course of a day, Vijay might implement his own code, review a subordinate's demo, teach a quick programming tutorial at an all-hands Lunch-and-Learn, interview a potential new hire, plan the next few months of product development with fellow Design members Travis and Tim, and walk a Contact team member through a particular feature built for a high-value client. It was a regular back-and-forth between putting out fires they hadn't anticipated and building a product that scaled with their business. When our first interview ended at 7, he turned right around and went back into the office.

The Division of Emotional Labor

As Beth and Vijay made clear, there are a lot of different roles and statuses in tech; it's not just a sea of developers tapping away at the keyboard. An important part of this story is that the work of producing that tech identity—training others to believe in the pivot—and the rewards from it are unevenly distributed across the firm, typically along gendered lines. Ji's startup Hearth was too young for this division of labor to reveal itself. InCrowd, in contrast, had grown rapidly since its founding in 2011, and by the time I arrived in 2014 it employed over thirty people, with more being added every month. Travis took an instrumental approach to these challenges, building the cultural and physical space of InCrowd with exactly this work ethic in mind, telling me "One of the things you have to accept is that people don't go to work these days, they go to sleepaway camp. The employer has to be more than a provider of salary but also an environment, continuing education, excitement." A shared culture was essential to securing buy-in for long hours and an unpredictable work style.

Women in "nontechnical" roles—whether formally included in their duties or not—were most often responsible for training their coworkers to embrace the uncertainty inherent to tech. Beth was clear that this sort of extra work was required: "You may have a role on paper but if all you're doing is what it says on that paper, then you're not doing your job." Beyond late-night conference calls or weekend reviews of potential new hires, this meant that time spent socializing at work and out of work became a research activity. She and the other Contact team members regularly coordinated brunches, concerts, hikes, and bar crawls for coworkers. It meant that her phone was filled not just with work emails, but Instagram photos, tweets, and Facebook posts from her coworkers.

Corporate identity became a social identity, an embrace of those community-building opportunities Travis mentioned alongside what Beth called "a mutual excitement for the product." She admitted that was cheesy, but that didn't make it less important. Beth drew on her friendships with members of other teams in order to put herself in the shoes of an engineer or salesperson and gain their buy-in on new projects. She also taught her team members to do the same. This instrumentalization of socialization by middle-management was not only encouraged by Travis but was one of the things InCrowd employees liked best about the company. That work friendships had a purpose was a sign that the company had a vision and that everything its employees did was bent toward that vision.

This vision was distinct from that of other big names in DC tech. When I asked how InCrowd compared to other DC tech stalwarts, many complained about companies and colleagues in tech that turned their office into a playground or let their workdays be taken over by drinking and video games. LivingSocial often played the heel in these stories: a high-profile startup lured to DC with hefty subsidies, but with a business model that wasn't helped by what InCrowders perceived as a party culture (Isaac and Benner 2015). Indeed, Beth preferred the InCrowd model to that of her previous startup, which "put these friendships first and that's why people were there."

Design was an all-male team led by Tim (who was Chinese American and nicknamed Captain Deploy) and Vijay (who was born in the United States but raised in Bangladesh and Britain), with largely South Asian subordinates except for Paul, who was White. Sales was split evenly between men and women and was led by Grayson (a White man Travis referred to as his second-in-command) and Suraj (an Iranian American, and the only non-White member of the Sales team). Contact was led by Beth (who was White) and was made up entirely of women of a mix of races, with the exception of three men—two White and one Black—two of whom exclusively focused on drawing up plans rather than on direct customer service. Sales secured customers, Contact maintained them, and Design almost never met them. Each team had its own office, though as InCrowd grew, it increasingly placed new employees in hallways or in the small corner spaces that used to be used for phone calls or one-on-one meetings. Travis tried to have a one-on-one with each employee every month.[3]

InCrowd's division of emotional labor had specific roles for specific divisions, each enrolled in the production of a hopeful tech culture. Design certainly had a role to play, even if other divisions joked about the team doing nothing but drinking Red Bull and building Connect Four bots for fun. Tim would post live updates to social networking site Twitter about demos Design carried out in its office, sharing InCrowd's business with the world. The clear glass wall around Design's office was always marked up, usually by Vijay, as the team diagrammed designs it was working on or brainstormed ideal features the team would love to work on in the future. The Wall of Awesome Ideas became a fixture of debates at company happy hour—every Friday at six—and of tours given to new employees and interns, prospective investors, or just other tech entrepreneurs visiting the office at Travis's invitation.[4]

But the bulk of this emotional work, the labor that led InCrowd employees to embrace the uncertainty of tech, was carried out by those in nontechnical roles. This happened formally, as part of the Contact and Sales teams' regular duties, and informally, as unpaid labor on the part of women InCrowders who kept the social life of the firm going.

The cheery tones of the Contact team were background music to the rest of InCrowd's work. Even as actual background music played softly from speakers in the middle of the office, you could hear the steady rhythm of their customer service scripts repeating over and over. Entry-level women working in Contact and Sales posted to the official company blog several times a week. This was the public face of the company and included everything from guides to popular Twitter hashtags used in the industry, to announcements of new features rolled out for customers, to core InCrowd values (hiring managers would be disappointed if interviewees had not researched these), to a breakdown of the annual InCrowd awards ceremony, where everyone received a funny certificate about their work personality. Contact members Jessica and Martine would regularly represent the firm at industry events or just fashionable DC parties, bringing InCrowd swag—t-shirts, toys, cards—with them. This was the face the firm showed to external stakeholders. Contact members kept that face up as part of their assigned duties.

But this work also happened internally, for the sake of their coworkers. Beyond coordinating InCrowd's social calendar and mediating between teams, the women in Contact also decorated the office in the company's orange and black colors, hung inspirational quotes, planned themed clothing days, and cheered the company on in their InCrowd t-shirts when it did public demos at DC tech events. They also brought a kiddie pool and coolers in to store beer for Friday happy hours, though men in Sales and Design were usually the ones going on beer runs. This work was not part of anyone's formally assigned duties. But without it, InCrowd wouldn't have felt like a startup, and its workers would not have bought into the flexibility the startup demanded of them.

Daily Pivots

The emotional labor that Beth and her colleagues undertook every day, and many nights, worked. InCrowd was a place where they not only wanted to work, but also embraced the extreme uncertainty and the daily pivots it required. When the company was much smaller, it relied on bootcamps like General Assembly and incubators like 1776 to provide employees with a startup mindset. That was what happened for Grayson, whom Travis praised for coming into the company at age twenty-three, ready to work around the clock and manage a team of his own and its rapidly shifting deadlines. By 2014, the vibrant company culture did that job for him. By then, Travis could joke with new hires about leaving behind the nine-to-five schedule that marked their old lives in government or nonprofits while he gave them a tour, and other employees would laugh along. Following Travis's example, his employees learned to internalize the demands of the company and its shifting fortunes. As Sales member Allen told an interviewee over the phone: "Every day is a school day, work hard and play hard. We are a startup but we all own part of the company, so we need to operate on the day-to-day a bit like how we want the company to operate."

"Every day is a school day" was a common refrain throughout DC tech, but especially at InCrowd. It signaled not just a commitment to lifelong training, but an openness to day-to-day changes in one's job—because there was always something new to learn, some new challenge to face. Where Ries and similar business gurus described the pivot primarily in terms of organizational strategies, the idea of embracing flexibility was important to InCrowd employees not just in terms of periodic gut-check moments for the firm as a whole but in terms of the changes individuals expected to see in their workdays. This meant not just a demanding schedule—long days, lots of different things to accomplish in them—but flexibility in how they organized their workplace and responded to incoming data.

While it might seem obvious, it's important to note that a laptop connected to cloud services is essential to the startup work style: meetings can be taken anywhere, ad hoc teams can assemble in different rooms or on the road. While "Every day is a school day" is a little more high-minded, "Can I show you something real quick?" is probably just as common in the day-to-day work of the company. Beyond that baseline, each team had different accessories to facilitate its own quickly changing tasks. Sales wore Bluetooth earpieces and business casual, always ready to take a call or meet a client. Contact wore a mix of business casual and the unofficial startup uniform of t-shirt and jeans. Each Contact member had a laptop screen open for active tasks and a larger monitor displaying ambient data for reference (e.g., billing information, standing instructions). Design dressed the most casually—its members rarely saw clients—and had setups similar to Contact, but the contents were different: usually command-line interfaces on the laptop and wireframes or the InCrowd website on the larger monitor. They had the only office with a TV on the wall. They would all turn to it once or twice a day for a demo.

InCrowd's office was flexible enough to accommodate new collaborations that emerged throughout the day. While long and narrow, it was a relatively open office. Individual employees had desks they usually used, teams had their own rooms with glass walls, and there were several unclaimed breakout rooms. That meant that these team dynamics would often spill over into adjacent rooms or hallways—Contact and Sales regularly took calls in closets—and this happened more and more as InCrowd grew. Travis and the team leaders kept the different teams abreast of each other, coordinating their different rhythms. Occasionally a big request from a large client or the rollout of updated software would force everyone to drop what they were doing, aligning their rhythms for the day.

These new rhythms were usually based on new data. Those data might be delivered by customers themselves, by InCrowd employees tracking their interaction with the product, by management tracking employees through the Salesforce software that recorded all client interactions, or by leadership's decisions about what new client sectors looked promising. Organizations like InCrowd are, in Neff and Stark's (2004) words, *permanently beta*: constantly testing new organizational forms because the firm's product and the infrastructure supporting its operations are both based on software that is itself constantly updated in response to user feedback and steady drips of performance data. This was especially clear during the B2B transition, but manifested too in the startups' growing pains as new roles were introduced that did not fit smoothly into the longstanding Sales-Design-Contact divisions: graphic designers, project managers, human resources, and so on.

It was also clear in the day-to-day work of the company, where riding the tide of constant feedback and performance data became a way to master the environment of extreme uncertainty. This was true both externally and internally. Contact team members regularly told clients that 75 percent of the InCrowd product was based on client feedback. This is a difficult thing to quantify, but the gist of it was surely true, for the simple reason that InCrowd needed to encourage more extensive and intensive use of its products to drive renewals and upsell existing clients on bigger software packages with additional user privileges or features (e.g., invoice management, client history pages).

Discussions within Contact about the problems a specific sector of clients—say, universities—faced would be relayed to Design and to Travis to prompt new technical solutions. Contact ensured a steady stream of client data by encouraging more active use of bells and whistles clients might not have noticed. Their activity could be tracked on InCrowd's end and used as a measure of a feature's success or failure. Sales would pick up new sorts of clients that drove new features, which created the demand for new client trainings. The feedback loop would also work in other directions, where the people supporting clients over time noticed new use cases that fed into new development, which encouraged Sales to go looking for more customers that fit the use case.

This permanently beta dynamic was also present internally. The customer relationship management software Salesforce tracked and aggregated every interaction an employee had with a client. This allowed for a birds-eye view of those relationships and made onboarding new employees, especially in Contact, faster because much of the "soft" work of relationship management was recorded, ready for review. InCrowd's customer database pinged every employee when a new client was secured, along with the amount of the sale and the commission the salesperson received. Beth praised this as a sign of an "open culture" where everyone was equally committed to the cause.

The pivot was thus not just an irregular crisis in the startup's life cycle, something that came up every year or two. InCrowd certainly went through a major gut-check moment as it became a B2B company. And it would experience several more moments like that, including when it was acquired several years later by a much-larger competitor for around $100 million. But there were many smaller pivots over the course of a day or month. Workers frequently changed tasks because incoming data changed, causing a cascade of other changes in roles, technologies, and office space. Buying into InCrowd's mission and the lifestyle surrounding it would seem to wreck work-life balance. And Beth did worry about this for younger employees especially. But on both anonymous satisfaction surveys and in my months at their office, InCrowd employees reported loving their jobs and everything came with them. They embraced the pivot. This empowered Travis's vision of his company: "If work is not part of your life or if you consider it two different things, then this is not the place for you. Because I want you to really believe in what you're doing and be really committed to the company."

#DCTech

Founders embraced the pivot as the defining feature of startup life because they had learned to identify with their firms. Startup workers embraced the pivot because it legitimated the long hours and quickly changing work practices required of them, even if the emotional work of teaching each other to embrace flexibility was not evenly distributed throughout the firm. Above the level of individual workers and firms, municipal elites in business and government positioned startups as a pivot for a city with an uncertain economic future.

Just as bridging organizations like 1776 or the various coding bootcamps taught individuals to pivot, they also built an identity for the sector as a whole in order to claim space in DC. This claim was built on tech's ability to help the city pivot away from government jobs, as well as decades of racialized poverty. The sector worked with government to redevelop the city in tech's flexible image. This included an activist cultural policy that legitimated the changes a flood of new, largely White professionals wrought on the city, as well as the city government's support of them. The DC tech community represented, to insiders and outsiders, the ideal type of worker and workplace—the end result of the access doctrine.

Mayor Vincent Gray, for example, never missed an opportunity to compare the influx of startup workers, and the real estate boom catering to them, with an older DC that was majority Black and, in his telling, crime-ridden, bereft of private investment, and dependent on the federal government for any job creation. "I could only think about what O Street used to be," he said at the Digital DC Thank You, held at the brand-new City Market on O condominium building in 2014. Gray then joked about how any apartment or hotel built here twenty years ago would have just been used for sex workers, before reviewing the benchmarks of his 2011–2016 economic development plan. The mayor himself thus participated in building up tech as a coherent sector, assisting the work companies like InCrowd conducted within their walls or the work organizations like 1776 did to bring entrepreneurs and startups across the city together.

Gray's economic development proposals contained the usual financial incentives to help kick-start incubators like 1776 or reduce taxes on passive income to incentivize venture capitalists, as well as a host of proposed changes that would improve the human capital stock of the city—both by reskilling the people already here and by incentivizing creatives to come and stay. The goal of his Creative Economy Strategy was one hundred thousand new jobs and a further $1 billion in tax revenue. Its cover depicted Black musicians alongside a majority-White cast of chefs, cameramen, and tech workers pointing hopefully at each other's screens. The creative economy included all these disparate jobs, linking them together through the ability to pivot: "In a Creative Economy, organizations are agile, continually offering new value to customers through innovations in products and delivery methods" (ODMPED 2014, 10). Pablum perhaps, but deployed in concert with very real economic development proposals and very real gentrification. As noted previously, and while there are of course many contributors, the rise of DC tech coincided with a demographic shift wherein Chocolate City lost its Black majority. What fruit did this bear for tech? As we've seen, networks of startups work with each other, related services (e.g., law, real estate, finance), and local government to create a local tech identity. They do this to attract a pool of talent (e.g., software developers who require little on-the-job training) multiple firms can draw from, to secure continued support from local and state governments, and to legitimate the social mission of a sector that is dominated by the White upper and middle classes and frequently indicted as an agent of gentrification and displacement (Stehlin

2016). In meetups and industry branding materials, this network was usually called DC Tech. The city government used other labels, enrolling the city as a whole in branding efforts such as Digital DC.

Building up a majority-White sector in a historically majority-Black city meant that leaders in tech and government needed to pose the influx of tech workers as a positive development for Black Washingtonians. It was a hard sell, but one Gray regularly tried to pitch at events like the March 2014 opening of the WeWork coworking space in Shaw. Douglas Development Corp. had long been seeking a tenant for the abandoned Wonder Bread factory it had gutted and renovated—one of the few remaining industrial spaces downtown. The space was overflowing by the time Mayor Gray arrived for the ribbon cutting downstairs, an empty warehouse space that would be leased to iStrategyLabs some months later. Black middle-schoolers from a STEM-focused charter school managed by Howard University wandered through in a line behind their teacher. They were the only signal of the historically Black university's presence nearby, the anchor of the neighborhood and one of the most important Black cultural institutions in the country.

"The District has come a long way," Gray announced. He ran down the list of successes from his tech-oriented economic development plan: growing the city's population from its historic low in the 1990s, bringing in more tax revenue, and moving it away from the federal government as the primary employer. Shout-outs were given to 1776's Challenge Cup competition, local startup TrackMaven's new founding round, and Blackboard founder Michael Chasen's foundational role in DC tech. They were helping the city pivot.

The mayor delivered a prize to the charter school students. Their technical accomplishments proved, he said, that they could go toe-to-toe with Sidwell Friends, the historically elite, majority-White private school to which Presidents Clinton and Obama had sent their children. Gray and WeWork CEO Adam Neumann stood on stage together to cut the ribbon, echoing the economic and social mission of tech, but also the personal one, with Neumann adding a motto that would resonate with InCrowd employees: "If you love what you do, you never have to work another day in your life."[5] Gray wore a tie, but every other CEO present—Travis included—went without.

A few weeks after the ribbon cutting, Travis and I sat down for our first formal interview, in one of InCrowd's breakout rooms. We spent our final twenty minutes together reviewing the financial incentives DC offered startups, and founders especially. He praised a since-canceled city program for paying the first $10,000 of a new hire's salary if they'd been unemployed in DC beforehand but were training for a new skill at their new employer.[6] "That is an awesome program! It does not dilute capital for entrepreneurs to put to work." But in general, he said the municipal government's incentives for startups to relocate to DC were "totally misaligned" with what founders needed, mere "lip service."

Regarding tax credits for hardware purchased in DC: "Who the fuck buys hardware in the store?" And: "This other thing where they have a real estate credit if you move to some shithole [the city-designated tech corridor along the gentrifying Seventh Street NW and Georgia Avenue NW strip]. No! I need to live here; I need to work here." Gray had unveiled that program at WeWork, and a few weeks later banners hung from streetlights along the strip: "Digital DC: The Innovation Capital." What Travis said he really needed were tax breaks, not so much for business expenses, because startups didn't make many big fixed-capital purchases, but from income and capital gains taxes so that he could save his profits and pump them back into InCrowd.

Despite working in the national Democratic party in the past, Travis confessed, now that he was a business owner, "I don't even know what government does. I just try to stay far from it." Still, I was a little taken aback when I asked if he thought the incentives package might change since Gray had lost the Democratic primary[7] to Muriel Bowser yesterday, and he raised an eyebrow, confused. He popped his head out of the conference room to shout down the hall, "Did anyone know there was an election yesterday?" and was met with shrugs and laughs. "Is she also African American?" he asked. I said she was and related a bit of her political history in Ward 4. Travis thanked me for the information and went about the interview. InCrowd had a mission; mayors would come and go. They would work around it.

DC was grounds for DC tech's various pivots, but not every part of the city could embrace the environment of radical uncertainty and transform itself in the same way. Both commercial and residential real estate developers were wholly on board, as were the bigger corporations represented by the DC Chamber of Commerce (motto: "Delivering the Capital"). They appreciated the influx of human capital and potential partners and acquisitions. The city's changing culture was a draw too. At industry events, I frequently met older investors and executives new to the tech scene, but unabashedly addicted to its culture. Some sections of city government were equally enthusiastic.

This love was not unconditional. Travis's ignoring the election was emblematic of the larger relationship between the sector and the city. DC tech needed the city's cultural cache and robust public infrastructure to attract talent and produce the agglomeration effects that characterize any growing sector (Audretsch 1998; Sassen 2001). The city needed DC tech to pivot away from the federal government and would pay for the privilege. But DC tech did not, of course, invest evenly across the city. Startup offices were overwhelmingly concentrated in wealthy, majority-White neighborhoods in Wards 1 and 2, with only a small percentage in other wards and none in the majority-Black neighborhoods east of the river (Taylor 2016).

Even the biggest players in DC tech were ready to leave, no matter the incentives on the table. Educational technology giant Blackboard booked around $700 million in annual revenue. After a long search for a new headquarters, in 2015 it announced a new 71,000 sq. ft. office in the city in which it was founded. The twelve-year lease was rewarded by the city with an annual subsidy of around $500,000. Those funds drew from DC's Qualified High Technology Company incentive program, which distributed around $40 million annually, largely to companies already in DC and without any dedicated agency to review the success of those or similar tax breaks (Bowser and DeWitt 2018). In 2019, only four years into its lease, Blackboard moved across the Potomac to Reston. The Virginia government did not provide any incentives, raising the question of whether DC's incentives mattered in the first place (Medici 2019; see also Jansa 2020).

The benefits of the city's pivot were unevenly distributed. But this new model of development, based in a citywide embrace of tech, forced other organizations, especially those concerned with poverty, to adopt the startup model of work and culture exemplified by InCrowd. Hoping to save themselves and the people they serve—both, the story goes, on the wrong side of the digital divide—places like schools and libraries attempted to pivot to a new identity and a new way of organizing their mission. They wanted a piece of tech's hope for themselves and the people they served. But older political commitments and harsh economic realities prevented a startup-style pivot and end up forcing impossible choices on these community spaces. This is the dynamic I call bootstrapping, and to see it in action we have to move a few blocks away from InCrowd's offices—to the MLK central library branch.

3 "More Than Just a Building to Sit In for the Day"

In 2004, DC was still awaiting the gentrification boom during which most of this book takes place. Mayor Anthony Williams had shepherded the city out of the austerity budgets imposed by the congressionally appointed Financial Control Board. Bond ratings had improved. Developers were planning. Williams spent much of the year negotiating his signature achievement: a gleaming new baseball stadium that would anchor redevelopment of the Southwest waterfront, for which the city eventually paid over $670 million, $135 million up front.

Most of the city was not seeing that sort of investment. There was much talk of radically downsizing agencies like the DC Public Library (DCPL) system. DCPL had been dealt a very public black eye in March 2004 when a computer virus took down every PC in the system for the entire month. Today, library officials talk about that as a wakeup call for the system, when they realized just how high the stakes were. Ultimately, the system crash was a symptom of austerity. In 1975, 620 full-time employees worked at twenty DCPL branches, but by 2004, only 431 worked in twenty-seven branches—nearly half as many full-time employees per branches. The library was, like most municipal services, understaffed and underfunded, its budget a much lower proportion of the municipal budget than in peer cities. This was borne of the budget cuts Williams insisted on as part of the plan to get financial control of the city back from Congress—an outsized manifestation of the majority-Black city versus majority-White state government conflicts that appear all over the United States. A *Washington Post* story about the computer outage said that DC libraries regularly opened late or closed early because of short staffing: "We've made do with very little for a long time without complaining," a librarian said. "The libraries are just crying out for help" (Fernandez 2004).

Another story later in the year explored the sorry state of library buildings—leaks, broken windows, overflowing garbage—that was caused, librarians said, by a mix of deferred maintenance and homeless Washingtonians seeking refuge.[1] One librarian at the Mt. Pleasant branch described her working conditions as "third world," saying, "There's just so much stuff that needs to be done and so little resources to do it with" (Murphy 2004). In response, Williams commissioned a Task Force on the Future of the District of Columbia Public Library System. Its 2005 report found that years of deferring maintenance and upgrades to technology and facilities had hobbled the system. It concluded, "The District of Columbia Public Library must be transformed if it is to be the successful, relevant institution needed and deserved by District residents. The transformation requires a new service dynamic as well as a new infrastructure of technology and facilities" (Williams 2005, 10). The library system, then, was short on political and financial support and overwhelmed by a homelessness crisis it was unequipped to handle. The month-long internet outage was the most visible sign of these tensions. It provoked a broad restructuring not just of the library's operations or facilities but of its identity. The new library would be based on new technologies and new service models. It would take a decade for this new library to appear.

In the spring of 2015, I walked into the Martin Luther King Jr. Memorial Library, the central branch of the DCPL system, for a meeting. Library police sat a desk inside the big glass doors facing G Street. A mural celebrating Dr. King's activism stood above the main foyer, overlooking a central help desk and a circulation desk against the opposite wall. Elevators and staircases took you up to offices, meeting rooms, the children's section, and more specialized collections. To the right of the main foyer was the Popular Services collection. To the left was the main computer lab, called the Digital Commons. Another help desk greeted you upon entering the Digital Commons, and from there you could either proceed on into the rows or computers and desks or turn right into the glass-walled presentation and coworking space called the Dream Lab.

I entered the Dream Lab, joining Dave, the mid-thirties White man at the head of MLK's digital programming, and Claire, a mid-forties Black woman and upper-level administrator at MLK. Most of the audience was made up of the Friends of the Library charity group. The Friends are middle- and upper-class White retirees who lobby the library on policy changes, fundraise, and run literacy classes and book drives. Dave delivered a presentation on the library's long-planned renovation. Our backs were to the glass cubicles separating the Dream Lab's presentation and coworking space from the Digital Commons, the 150 seats of which were full, as usual, and dominated by the city's homeless population—mostly older Black patrons, more men than women, who walked over to MLK every day if they weren't dropped off out front by the shelter shuttles that also did pick-up runs in the evening.

Dave, eyes gleaming, asked if we'd like a tour of the new Fab Lab makerspace upstairs—a reclaimed meeting room intended as a preview of the fruits the renovation would bear. So we walked past the help desk where a librarian monitored the whirring 3-D printer, through the main foyer where a mural of Dr. King overlooked members of DC Tech setting up hundreds of chairs for their monthly demo series, up two floors on the elevator, past one of the video visitation rooms for DC Jail, around the corner from the Black studies collection, back into the cavernous stairwell that had been a gay cruising spot in the 1980s, through some locked double doors, and into a sunny meeting room with floor-to-ceiling windows that looked out onto the Morton's the Steakhouse next door.

Dave regaled us his vision for a new kind of library. It was hard not to get caught up in his enthusiasm for the 3-D printers, the laser cutters, the CNC fabrication machine, and the scattered laptops. He pitched the maker skills the Fab Lab would teach as a new literacy for the information economy, something that could help defeat the STEM gap and provide the creative, technical workers he said we were so desperately short on. Consumers would learn to maintain their devices and save the environment. Skilled technologists would have a new space to inspire underprivileged communities. One Friend pitched it as a poetry lab to upgrade the arts for the twenty-first century. Dave said they were "testing for tomorrow," a tomorrow where people could say, "I learned to code at the library, I got a job because of the course I took at the library." Dave described this lab—and the others that would soon emerge—as "wild extensions" of the Dream Lab downstairs. This meant not just a new suite of tools, but new modes of librarianship, new ways of understanding library patrons, and, fundamentally, a new library. It was exactly what the task force on DCPL's future had called for a decade prior.

There was so much hope in the Fab Lab. Much of it recycled from three-year-old promises about the power of Digital Commons on the ground floor, where most patrons spent most of their time. That space was a massive upgrade from the fourteen Dells that had previously made up the main computer lab of the central library branch of the nation's capital. There was so much pressure placed on the Fab Lab and the technologies and librarians inside it—even though, from its opening until MLK closed for renovations in 2017, the room was mostly used by library visitors rather than the homeless patrons that were there all day every day. The access doctrine flourished in the library. The new labs promised a brighter future for those with the right tools and skills. But the organizational restructuring the access doctrine prompted did not help everyone in the library flourish.

These upgrades project a reassuring vision of the future in a city where a post-2008 flood of new tech workers was accompanied by housing and jobs crises. Between 2000 and 2015, DC's affordable housing stock decreased by 50 percent (Rivers 2015), and between 2011 and 2014 there was a 12 percent increase in total homelessness and a 29 percent increase in the number of homeless families (HUD 2014). In that time, MLK served an important role as a safe space for homeless Washingtonians with nowhere else to go during the day. But the library was poorly equipped to handle this role. It did not have the necessary resources to support those in crisis—whether single-occupancy restrooms for folks who needed a space to groom, or dedicated spaces to sleep, or the necessary training to aid folks with serious and persistent mental illness. Moreover, creating these resources would not attract the sort of support the library needed to secure its own future.

Whereas DC startups represented the "right" side of the digital divide, Gore's "information haves," the public library served the "information have-nots." At InCrowd, we saw how startups and the people within them embraced the radical uncertainty of their sector and learned to pivot—both as a large-scale matter of shifting corporate strategies and a day-to-day matter of shifting tasks. That is the ideal type of organization and work style within the access doctrine. Institutions like MLK, or public libraries more generally, hope to teach their patrons to survive the information economy in this style. But to do so, the library must transform itself—literally, in MLK's case. The library changed to help information economy stragglers change in turn. This process enrolled a host of materials at a variety of scales: from boxy black Dell OptiPlex 755 PCs and shiny new MakerBots, to a historic public space in the middle of the DowntownDC Business Improvement District, to a city government desperate to project an urban identity independent of the federal government, to the mass of patrons living through the day-to-day reality of racialized poverty.

This chapter explores the first of two institutions that take the access doctrine as their mission. Libraries and, as the next chapter explores, charter schools that embrace the access doctrine begin a process of bootstrapping: a series of stark organizational reforms that reorient the institution around the idea that poverty can be overcome with the right tools and the right skills. This changes the sort of care the institution provides. At MLK, this process focused on a new culture of digital professionalism. Librarians were reimagined as knowledge workers, rather than helping professionals. They were tasked with training patrons who were reimagined as potential knowledge workers. The physical space was redesigned to support this process.

The language, work styles, and organizational cultures of startups are the ideal organizational type for the library—but libraries can never be startups. Their resource base is different; their mission is different. Beyond those structural limits, there is the fact that patrons and librarians themselves resist these changes, building their own libraries in the corners of a computer lab or between the stacks. Such resistance persists, but it cannot ultimately stop the library being remade around an ideal of digital professionalism because that process simplifies the library's complex and competing set of priorities in the present and guarantees political legitimacy and material resources for the future.

Most librarians, and of course patrons, understood how complex the problem of homelessness was. Most wanted to do what they could to solve the problem or at least assuage its pain. But the solutions they could pursue were constrained by the access doctrine. For the library to maintain this hope in "using the technology to improve their lives," as librarian Grant put it, it must necessarily regulate or eliminate competing uses for the library space. To investigate this process, I first explore how the library's bootstrapping reorganizes its space around digital professionalism, and then how patrons adapt to this process and produce other versions of the space to meet their own needs.

A Transformational Library

The library's bootstrapping process generated a conflict over what the library was *for*. Everyone agreed that the library was a critical digital resource, but the nature and purpose of that resource were contested. Were the Digital

Commons[2] and connected spaces refuges, some of the last remaining public spaces in the city where anyone could rest for the day without purchasing anything? Or were they training centers, spaces that could provide the skills and tools that would help people move out of the library and into the jobs of the future? To resolve this conflict, the library and librarians dictated the correct way to use the space and the technologies in it: not just the desktop computers, but the new 3-D printers; the Memory Lab personal archiving suite; the phones, tablets, and laptops people bring with them; the desks and walls; the music or drugs that found their way in; people's clothes, bags, and, of course, books. To make the right space, everything needed to be put in its place.

The work of present-day reorganization was determined by a future vision for long-overdue renovation to the central branch. Finalized in 2014, at a projected cost of $208 million, this vision was repeatedly articulated at the highest levels of DCPL and local government. The mayor's office and new Chief Librarian Richard Reyes-Gavilan worked hard to advertise the work architectural firms Mecanoo Architecten and Martinez + Johnson[3] would do. Their renovation would double down on recent digital upgrades. It would also move MLK's cubicles and stairways from a closed, transactional space, wherein librarians provided patrons with items, to an open, transformational space that would offer learning and training opportunities to the whole city. In day-to-day operations, professionalizing the library space meant promoting the use of information technology that, at least in appearance, offered skills-training opportunities. Disruptions to that process had to be regulated or eliminated.

What did that regulation look like? As April, a librarian in her mid-twenties, patrolled her branch with colleagues, she gave out imaginary stickers to patrons they thought were using the space appropriately ("gold star if you manage to use the library appropriately..."), inappropriately ("special snowflake if you really think the rules don't apply to you..."), or just wrong ("paint bucket for 'You're as dumb as paint.' You're teachable, you're just dumb"). They walked the stacks and the computer lab, giving out stickers whenever they saw patrons engaged in self-talk, fighting with each other, eating, watching porn, touching themselves or a partner, or bedding down for a nap on a strip of cardboard in the reference section.

April had a master's degree in library and information science. She was a middle-class White woman who had moved to DC for a secure but stressful job. She could tell you how to verify Google results, do basic HTML, and find your nearest polling station at election time. She loved open access and President Obama. She and her coworkers understood themselves as helpers. That professional identity was necessarily shaped by their relationships with the people they were tasked with helping, who lacked those skills and whose comparative lack of freedom in the city was marked by race and class. April served patrons who were poor or working-class, who had only a high school diploma (if that), who were much younger or much older, who were Black and Latinx, who were priced out of DC housing, who were living with mental illness, who mistook socialsecurity.com for socialsecurity.gov.

These were April's patrons, or *customers* as she and most newer librarians called them. And while she ostensibly served them, her sticker system showed there was a power imbalance in that dynamic; help largely proceeded on the helpers' terms. Indeed, Shawn, the "computer man" from the introduction, explicitly compared librarians' labor to that of the social workers he had met in different social service agencies: sitting down to chat with regulars, acting as a reference for jobs, actively looking for opportunities to help. They were helping professionals, but there is a line between librarian and patron, just like there is between student and teacher, nurse and patient, social worker and client.

Like the public school or the clinic, American libraries provide care, but on terms of self-improvement. This has been true since at least the founding of the American Library Association in 1893. Most of the librarians I encountered described their profession in classed and gendered terms as a *pink-collar* one. April called them "mavens of knowledge" (Fox and Olson 2013). It is a long tradition. White middle-class women in the progressive era worked as readers' advisors, teaching immigrant patrons to move away from entertainment materials and toward Anglo-American classics, inculcating sufficient literacy to enter formal job and housing markets (Garrison 1979; Luyt 2001; Wiegand 1986). This outreach took on renewed importance when, as chapter 1 showed, the neoliberal revolution gutted the welfare state and positioned the networked opportunities of the internet as a substitute. The new regulatory regime introduced by the Clinton administration gave the United States some of the slowest, most expensive internet in the developed world and made community access centers like schools and libraries the only places to get it for free.

Libraries, of course, do much more than digital professionalization. On any given day, the library might be assisting with health insurance enrollments, providing story time or book clubs for multiple ages and languages, offering yoga classes or film screenings, presenting an exhibit on the history of punk music in DC, or helping patrons file their tax returns. Different DCPL branches take on different tasks that fit their neighborhoods. Many branches were recently redesigned themselves, some with new glass facades designed by architects at the Freelon Group—who lost out on the MLK contract. The transparent buildings were meant to better connect the space to the community. Branch librarians told me they were proud of these gorgeous new angular buildings. Still, many preferred the old cement blocks with clear sight lines down each row of books. Staffing shortages meant a few people were responsible for a lot of space at any one time.

The long-planned renovation of MLK loomed over my time in the library, shaping the work of librarians like April and Grant and the routines of patrons like Shawn and Mia. Every librarian said it was overdue, calling the interior of the building dark, uninviting, and difficult to navigate. Even though they were nervous about their ability to control the outcome, they evangelized the renovation. It became part of their job—to the point that that many wore "Ask me about the renovation!" buttons during the public comment period.

As in startups, this culture grew thicker the higher you went in the org chart. Executive Director Reyes-Gavilan was recruited to lead DCPL, and at this moment that meant completing the renovation in a way that would emphasize MLK's centrality not just to the branch system, but to the changing city. In 2015, I attended a forum at the Hamilton gastropub where he spoke about the renovation with police, the Advisory Neighborhood Commission that oversaw development in the area around MLK,[4] and local business owners. These were the neighborhood power brokers, and any big change in the neighborhood had to be responsive to their needs.

Reyes-Gavilan was blunt: "It was outdated the day it opened. It has been an unloved structure for a long time. But it has always had potential." That potential was the focus whenever he showed off architectural renderings: a new library with expanded labs, space for startups, a rooftop garden, and what appeared to be a TED Talk in the central foyer. So though there was certainly much more going on at the library than digital professionalization, those were the activities that connected most directly to its bright future. And that's why it became more of a focus over time.

This dynamic is not unique to DCPL. Stevenson (2011) shows that US and Canadian libraries were a key site of welfare state investment in the

1950s and 1960s, and libraries' description of their public service mission emphasized a well-credentialed workforce ready to help with anything. In the 1990s, that began to shift as customers and information technology became the focus of library discourse, rather than librarians, partly as an attempt to justify the library mission in the face of broad cuts to welfare state services. Libraries and library schools began to describe librarianship as yet another information economy profession whose product mainly consisted of serving customers through information technology. The work of care becomes the work of technology provision and skills training.

MLK's librarians were thus of a class with the digital professionals, government officials, and business owners filling those rooms to hear about the library's future in the Dream Lab, before Dave showed us a preview of that future in the Fab Lab. In contrast, the crowd sitting on the other side of those glass walls in the Digital Commons were the lifeblood of the library's present, but not necessarily included in its future. They were largely Black, where the future-planners were largely White. They weren't wearing suits, and though smartphones were common, they were rarely the new iPhones you saw more frequently in the Dream Lab or other meeting spaces. Not that they were deviceless. Shawn was not unusual among my homeless informants for having two phones—a federally subsidized TracFone for calls and a Sprint smartphone he bought himself for email, texts, music, and social media. Tablets were common, laptops less so.

Librarians certainly tried to include patrons in the planning process, but there were limits. Obscure architectural and bureaucratic language made collaboration difficult. Each meeting referenced a longer narrative of prior meetings. And they were usually held in the early evening, when office workers had just left their jobs, but homeless patrons were often seeking out church dinners or shelter buses.

The institution bootstrapped to save itself and serve its patrons, but that process was not equally responsive to everyone and thus patrons who were not already digital professionals were less able to shape the library's future. As far as the planning process was imagined, one group was there to help and another there to be helped. And while the day-to-day work of the library was far more complicated than that, this way of planning the future began to structure that work. The access doctrine framed the library as a community access center, where patrons could find—through new tools or skills—new opportunities for competition. Although at the end of the day, most patrons I met came to MLK not for a specific digital opportunity but simply because they had nowhere else to spend their day.

The library's future depended on its ability to bootstrap and become a professional space that trained future digital professionals. Patrons, occasionally with staff support, still carved out different kinds of public space within MLK. But the library overpowered those efforts as the renovation neared. It had to. Reshaping the library's space and operations around digital professionalism proved MLK's value to the funders, politicians, and community power brokers who could make the renovation happen. It was the library's way of showing outsiders that it was an essential piece of DC's information economy, an institution that could help "information have-nots" cross the digital divide.

Inside the organization, however, patrons had many needs—homelessness, illness, joblessness—that the library faced because of its role as one of the last public spaces in the city, needs it wasn't equipped to meet. In the long term, the access doctrine likely worsened this overload by contributing to a general political project of defunding the welfare state while adding digital resource provisioning to the library's long list of duties. But in terms of daily operations, the access doctrine refined and focused librarians' jobs, reducing the complex problem of urban poverty to a much more basic binary: a digital divide that could be crossed with the right tools and skills.

Professionalizing the Library

Personal computing was thus the terrain on which the fight for MLK's identity was fought, where the bootstrapping process unfolded. There are many things one can do with a PC, and many things one can do at the library. At MLK, personal computing was bent toward the professional norms of digital professionalism. These values were at times explicit and other times less so, embedded in library infrastructure, lessons taught about personal computing, and the selection criteria for new librarians. This helped the library and librarians on the ground manage the overwhelming, often competing demands placed on the institution: at once research space, community center, media repository, day shelter, local archive, meeting place, training facility, government point of service, and much, much more. Organizing the library as a space for digital professionals provided a way to manage these diverse needs, and this held great appeal for stressed librarians trying to keep the place running or worried patrons trying to find a place in it.

Nowhere was this more visible than in the simple process of managing demand for the library's computers, the system by which MLK quite literally arranged and prioritized the many people claiming a piece of its most public space: the gray rows of the Digital Commons computer lab. In the Digital Commons, patrons had a fair amount of freedom to do as they chose with the available desktop PCs, let alone the tables at the back ready for their own phones, tablets, and laptops. There, librarians were less able to control patrons' computing as compared to the instructional rooms on the upper floors. The three or four librarians on duty in the Digital Commons could not possibly keep an eye on everyone's screen, even if they were assisted by the armed representatives of the DC Public Library Police, although the division of labor between the two sets of public employees largely dictated that police worried more about behavior and less about library materials. Some regulation was necessarily delegated to the library's digital infrastructure, such as the Pharos queue system and the PCs' internet filters.[5]

Patrons used their library cards to sign up for sessions at a central terminal by the printer. They were then directed to a queue displayed on a pair of large, wall-mounted screens to wait to be assigned to one of the seventy-four desktop machines of the Digital Commons. Patrons could not log into a computer to which they were not assigned.[6]

In 2012, Elena, a mid-twenties librarian who supervised the three-hour waits for the fourteen computers in the old Popular Services computer lab, told me that even triple the number of computers would not be enough to meet patron demand. She was right, especially so during DC's sweltering summers. Then, unlike winter, there was no right-to-shelter ordinance for the homeless. Patron Mia told me that her shelter let her hang out all day during hypothermia season, but not during the summer. As the mercury rose, the wait for a PC could extend to over an hour. Severe weather, hot or cold, meant more crowds and longer waits. I entered the library as it opened on a two-hour delay in February 2014, a day after a foot of snow hit, and a fifteen-person line immediately formed at the sign-up desk and didn't subside for three hours. Within forty-five minutes after opening, every seat was filled.

The queue for computers in the Digital Commons indexed the strain placed on MLK. That strain fell on the computer lab both because that's what patrons wanted and because there was no other space to accommodate that many people in the building during the normal course of a day. When demand was high, librarians were quite strict about keeping the queue moving. Elena had no sympathy for those who signed up but missed their alert for a cigarette break or something else: "You know how this works. You know the rules. You missed your turn. Too bad." Pharos also allowed librarians to monitor every session's activity from a central terminal and choose to end or extend the seventy-minute session. Patrons watching porn repeatedly might find a pop-up screen saying, "Please don't do that," a privilege librarian Rachel frequently exercised against patrons whom she felt were using the internet, and by extension the library, incorrectly. A later upgrade to EnvisionWare software added a behavior code to which patrons had to agree before logging on. Patrons chafed at this surveillance, part of a wider network that includes a dozen cameras in the computer lab alone, positioned above Wi-Fi access points, and constant patrols by librarians and police.

Mia was quick to complain about this surveillance network, especially Segway-riding police patrols, but was largely resigned to it. The library after all is only one of the government offices that those in the shelter system regularly visit, all of which demand consent to regular surveillance (Eubanks 2006). Having a librarian note your internet activity is unexceptional when you also have your diet and sleep schedule policed at the group house, your sexual activities and social life critiqued by clinicians, and your daily purchases scrutinized when applying for food stamps or housing assistance. "The system is designed to break you down physically and mentally," Mia said. "When you've been in this situation for a while you realize very early on that your life is very public."

Librarians did not only restrict unprofessional technology use in the Digital Commons. They also actively encouraged professional uses. Patrons working on a job application or filing for unemployment insurance with the municipal government could raise their hand, have a librarian walk over and check in, request extra time, and usually have an additional session tacked on. These distinctions between correct and incorrect computer use did not appear in the library's posted rules for computer use, besides the boilerplate notice that inappropriate materials will be filtered, but every librarian I spoke with admitted to acting on them, and every patron I spoke with admitted to having the "rules" explained to them at one point or another.

These moments of professionalization were more explicit in the complex hierarchy of PC classes across MLK. Classes for beginners (e.g., introductions to email, Microsoft Word, or PC basics) took place during the day, in a third-floor classroom with about forty PCs, away from the bustle of the Dream Lab and Digital Commons on the ground floor. Attendees were mostly older Black men and women without their own laptops. They were trying to reenter the labor market, upgrade from their current low-level position in the service economy, or learn the skills necessary to communicate with friends and family—often abroad—or manage interactions with social services.

Classes for more skilled students ranged from Adobe Creative Suite to Python to mapping sessions with Mapbox. Mapbox was a startup the library granted free workspace to in the early days of the Dream Lab in exchange for teaching classes. In June 2015, it raised $52.55 million in Series B investment. It was a regular talking point of library administrators, a clear success for the Dream Lab in general and its outreach efforts with tech startups in particular. Classes like Mapbox's, labeled intermediate or advanced, mostly took place at night in the Dream Lab. Those students were required to bring their own laptops. That crowd was younger, Whiter—a larger relative proportion of White patrons compared to those present earlier in the day—and dressed in the clothes they just left the office wearing.

There were several sessions of Intro to PC Basics upstairs every week. Many were taught by Betsy, a middle-aged Black librarian who encouraged her students to repeat these foundational lessons until they felt confident, gently ribbing them all the while. "This is for folks who have no clue and that might be you!" Her class emphasized beginner skills, like how to right- and left-click or create folders, but also concepts: the different names for a flash drive or hard drive, the logic of file trees or deletion, the "proper language of the industry" that prevented embarrassment at a job interview. She often referenced the civil service exam—even though there was no longer a single exam and most students were not applying to those jobs still requiring one.

Independent, PC-based office work was not only a story that drove Betsy's classes. It was a model built into her exercises and instructions: reciting the technical terms for different pieces of hardware to get students past "whatchamacallit," the typing motions that Betsy differentiated from those used on typewriters in the old civil service exams, the confidence to not request help but to close a program and reset the computer if the anticipated caption does not pop up. Students might apply these skills to a variety of domains, but for the library, the arc of personal computing bends toward professionalization. After questions on demographics, skills, and the instructor, the DCPL postclass survey that all students were asked to complete included ten items in response to the question, "How will this class impact your life?" Four, including the first three, explicitly addressed professionalization: job performance, creating professional documents, job search, and online business presence and marketing.

This professionalizing mission emerged in part from intraorganizational pressures that come from leaders or from staff's management of competing, everyday priorities. Those leaders were often responding, as Reyes-Gavilan showed, to a larger set of incentives that shaped the whole organizational field of librarianship: the funding and legitimacy offered by philanthropies and local governments. Other changes in the librarian profession pushed librarians further down the organizational hierarchy to adopt a professionalizing mission not necessarily because of a specific incentive but because a clear set of best practices provided useful guidance in an environment of extreme uncertainty. A focus on digital professionalism emerged from both credentialing organizations that regulated the profession and local organizational changes that fit the profession to MLK's specific needs.

The master of library science degree (MLS, or sometimes the MLIS with the addition of *information*) is a prerequisite credential for promotion or administrative duties. In recent decades, librarians have begun receiving that degree not from traditional library schools but their successors: information schools (Olson and Grudin 2009). In interviews, veteran librarians often regretted this transformation, explaining that an older public service culture had been replaced with a more professional, technical one.

Becca, a librarian in her late thirties who "can't imagine doing anything else," was training for her MLS in 2000 at the University of Maryland when the College of Library and Information Services changed its name to the College of Information Studies. She read the shift as the tragic downfall of the profession, an embrace of technical over service values: "Man there was a stink like you would not believe. You're going to eliminate "libraries" first of all and then you're going from Service to Science.[7] Leaving the people out, that face-to-face. Nothing wrong with theory, I love theory, but people are somehow getting kicked out.... There was a big, big stink about the person-to-person service versus the cold, electronic seemingly end-all approach that looked at face-to-face as kind of antiquated. No!" Becca's own career arc provided ample evidence of the pressures placed on the profession. At one point, state budget cuts closed the medical library at which she was working. She was forced to reapply for her old job as a new part-time position and rebuild the print collection in a digital format. Credentialed right when the information school (iSchool) movement really took off, she was the most junior librarian I interviewed who consistently called her patrons *patrons* rather than *customers*.

Contrary to Becca's description, this was not a total loss of a public service culture but a sign of how the access doctrine pushed that culture to understand service as skills training and technology provision. The shift is encouraged from above by state institutions and major grant-making bodies, such as the Bill & Melinda Gates Foundation, Google, and the Institute of Museum and Library Services (IMLS). Pitching to external funders became a matter of survival for twenty-first-century libraries that found their traditional funding streams—largely local taxes—gutted during the recession (Kernochan 2016; Schatteman and Bingle 2015).[8] Between 2004 and 2016, total revenues for US libraries fell 21 percent. The steepest drop came after the recession, and most state library networks did not recover (IMLS 2017). In DC, that meant cuts to service hours and staffing in the immediate wake of the 2008 crash that took years to restore, even as new branches were opening.

Victims of welfare state retrenchment, public libraries adopted a new mission that legitimated their existence in an economy defined by uncertainty and positioned them to help their patrons survive the same. Stevenson and Domsy (2016) found that a similar discursive shift in how Canadian libraries explained their mission prompted major technological and staffing shifts: technical services staff shrank as collections shrank or arrived shelf-ready; greater automation of reference and circulation functions meant reference librarians were often replaced by paraprofessionals without job-specific training.

Librarians embraced the access doctrine not just because of the overall environment of fiscal austerity or because of changes in their credentialing institutions, but because their bootstrapping libraries changed their hiring criteria. Hiring the right kind of librarians was a way of enforcing the right way to use the library. Eugene, a mid-twenties librarian, explained to me that the Digital Commons' technology suite was incomplete without librarians to match. The Adobe Creative Suites, the 3-D printer and Espresso Book Machine, the Dream Lab's glass conference rooms loaned out to local startups: these all required a group of librarians who were younger, hipper,

Whiter, and more digitally literate than the branch's veterans. Few held an MLS. Previous chief librarian Ginnie Cooper's administration had laid off a group of long-tenured, Black MLK veterans before they could collect their pensions—a case their union was still pursuing—and replaced them with ten majority-White members of what Eugene called "the hipster contingent": "It really looked like 'We're going to hire young, hip people.' I'm the library's idea of 'hip' which is sad. And that was to staff Digital Commons. Similar things have also been happening out at branches. ... I think the people who really would have had a lot of problems with that, you know starting from point one and just fighting it all the way, they've been gotten rid of."

Their t-shirts, jeans, dyed hair, and informal service style—pulling up a chair to chat up regulars, organizing basement hackathons with new library-approved drones—often made me do a double-take on evenings when I arrived at MLK after spending the day at InCrowd's office. The two sets of employees were indistinguishable at first glance. This enthusiastic startup aesthetic was essential to producing the Digital Commons space. Deliberately or not, the "hipster contingent"—cool, young, skilled White people in t-shirts and jeans—very much embodied the hope linking personal computing with social mobility. They were the access doctrine made flesh. The library pivoted to them, and their job was to help patrons pivot in turn. Their labor was not only the technical work of helping to set up resumes or recover email passwords, but the work of projecting that hope, of performing the future of the public library.

Keeping the Library Unprofessional

The library applied tremendous energy and resources—best symbolized by the hiring of the hipster contingent and the $208 million renovation budget—to become a hopeful, professionalizing space. Reyes-Gavilan evangelized this shift from the top down, and it only gained momentum as the renovation drew closer. The Dutch architecture firm that designed the new MLK even produced a documentary about the renovation that linked King's political values with the design philosophy of Ludwig Mies van der Rohe, the Dutch modernist whose famous black metal facade made the building an historic site—which also made renovations more difficult to plan and approve. The trailer shows architects pushing past "scary" interior doors and lifting rusted blinds out of their way as they discuss how they're going to bring people, instead of books, into the light. Their conversations carry on as the camera pans through librarians' cubicles and, again and again, the rows of computer users in the Digital Commons.

This story leads to a particular, professionalizing way of using the library and its digital resources. It is individualized through long rows of PCs or desks with plugs. It is transparent with glass cubicles and open air. It is surveilled to orient patrons toward the habits and methods of office work. But the institution's production of space is always, to greater or lesser degrees, resisted by the people within. Many homeless patrons, with whom I spent the majority of my time in MLK, recognized that the institution and the people working in it could never wholly commit to the neoliberal access doctrine, in part because they still valued the older, liberal ideal of the library as a space of public service.

There was thus an ongoing debate among the majority-White professionals over what the library of the future owed to the library of the past. Patrons picked up on this, strategically using different librarians' sympathies for different positions or, in other cases, the very indeterminacy of the debate to claim space in the library for their own purposes. Patrons may not have been among the stakeholders shaping the bootstrapping process—it was supposed to serve them, but they could not offer the organization resources or legitimacy—but it was a contentious process. And as it unfolded, patrons could take advantage of it in order to adapt the bootstrapping library to their needs or to build alternative spaces of their own—libraries within the library. That the new, professionalizing library did not fit some patrons' needs did not stop patrons or sympathetic librarians from using the symbols and technologies associated with whiteness and professionalism to secure aid, comfort, and community (Brock 2020).[9]

Adapting to the Bootstrapping Library

Patrons knew which librarians were best able to help them with particular tasks, such as filling out forms for food stamps, affordable housing, and the like. Patrons knew this and picked particular librarians with good reputations for particular tasks. These adaptations to the library space relied on an intimate understanding of an important and unresolved ideological tension within most librarians, an uneasy gap between two ideal libraries: a professionalizing space full of future entrepreneurs and a public space full of citizens to be served. This was easiest to see when pornography came up.

Most patrons acknowledged that watching pornography was "doing the library wrong"—most of it was filtered, after all—but knew that they could get away with it anyway with a little work: choosing the right site that the filter had yet to catch, switching between windows so librarians patrolling the rows didn't catch them. It was a very rare day that I did not spot several screens of hardcore pornography in Digital Commons' open rows of PCs.

As Rachel, an early-thirties librarian, explained, she and her peers wanted to preserve the professionalism of the community access center. Watching porn at the office, of course, is generally frowned upon. But they also wanted to preserve the library as a public space, what they understood as their profession's historical legacy. Rachel couched the issue of internet porn in a series of contingencies: "If you look at a nudie picture and you do it in a way that other people don't have to see it, it doesn't bother me. ... If they don't have access to a computer in another spot and it's an outlet, it doesn't really bother me." Her branch's lab was not laid out in MLK's neat rows, so she could turn a blind eye when patrons opened porn on more isolated computers.

Elena echoed her, telling me internet porn was "morally awkward and professionally awkward ... an unclear gray area." This conflict between what they had to do and what they wanted to do extended to other areas, but porn was the first example of doing the library wrong that everyone I spoke with jumped to, just as job applications were the go-to example of "doing the library right." Despite this consistent rhetorical positioning of porn as a *negative* example of how to use the library, it was clear that porn was a *positive* example of the sort of service librarians believed they should provide: giving people the space and the materials they could not get elsewhere. Librarians felt conflicted about porn because this particular digital resource exposed the conflict between two distinct but overlapping institutional cultures: a public service library that welcomed all comers and a bootstrapping library that trained digital professionals.

But librarians were not the only municipal employees enforcing the rules of the space. There was also the custodial staff and a heavy police presence—especially at MLK, where five or six officers were on duty at a time. Members of a dedicated library police force separate from the Metropolitan Police Department, they roamed the Digital Commons, hands on holstered pistols, walkie-talkies the loudest thing in a quiet room. They had a desk at the entrance and a control room upstairs to review the surveillance camera network. They were allowed to touch patrons, where librarians were not. Police tended to enforce norms for sleeping, drugs, fights, phones, theft, or exposure, rather than personal computing proper—unless a librarian called them in to act as the stern right hand, enforcing the liberal left hand's rules.[10] Adapting to the police required less negotiation and more stealth.

Any day Mia and the friends she called *the crew* were not at a day program for a clinic or a visit with social services (which was often; being poor and "in the system" is time-consuming and expensive), they were at the library. But they only began their regular routine at MLK in late 2013. The crew had moved from branch to branch, fleeing police who hassled them, Shawn especially, for sleeping at computer desks or speaking too loudly on the phone. A year later, Shawn and his girlfriend Ebony were still frightened when an officer they recognized from the Northwest One branch visited MLK. At the smaller branches with only a dozen or so PCs, it is easier to spot patrons using the library incorrectly. MLK's greater size allowed for greater anonymity. But Mia was still frequently tapped by police when she looked like she was dozing off, when in fact her astigmatism forced her to lean in close to her laptop screen.

Despite this general animosity, Mia and Ebony, like many homeless patrons, also strategically leveraged their relationships with individual police officers. Food was forbidden in the Digital Commons, but Mia was diabetic and needed regular snacks. So she waited to eat until specific police officers were on patrol and hid the food when unfriendly ones were roaming the lab. She was strategic but undaunted: "The ones that are assholes better stay the hell away from me. I know that. They know I'll chew them out in a quick heartbeat." In the fall of 2014, Mia supported Ebony in seeking help from the police after her assault by a group of teenage girls in the bathroom. The police brought the two of them back into their surveillance room, a rare look behind the scenes, and asked for descriptions as they scanned the library for the assailants.

Beyond adapting to library personnel, patrons also learned to adapt to organizational infrastructure, developing a slew of strategies to manage the login system and the queue. "You only get two [computer sessions] a day," Mia said. She went on, explaining how she adapted to these limits: "If you're diligently looking for work, trying to update your resume, looking for different training programs, looking for housing … if you're in my situation and you're trying to do all of those things in a day, an hour and ten minutes [per session] is not enough. After your two, that's it. You got to scramble around and be like 'Hey are you using your other reservation? Can I use your library card?' Or ask a family member to make a library card even though they never go to the library. For a while, I was using my uncle's library card because he never goes. He has it but just never uses it. I was going to ask my grandma to make one as well because she's a movie person." It was a great deal of work to actually carry out these adaptations—whether that was work that patrons did ahead of time, like Mia did, or work that happened in the moment, as a session was running down. Ebony, before she was gifted a used laptop by an older friend from her church, would email whatever she was working on to herself before her session ended, run back and grab Mia's library card, and book a new session as quickly as possible.

Other patrons would work in pairs, hoping the police did not spot them and enforce the one-user-per-computer rule. One partner would sign up for a session where they collaborated on a task, and once the countdown timer appeared onscreen the other would run to the queue to reserve a new machine where they could continue the task. Credential-swapping as a form of mutual aid was of course not limited to the library. Patrons shared other state-issued ID cards—SNAP cards or clinic-issued Metro passes—when and where they felt they would not get caught. They took this practice into the bootstrapping library to make the space work for them.

If the queue indexed the strain placed on the neoliberal library and demonstrated how it managed that strain, queue gaming, porn watching, and police dodging showed that a space of public service, fit for whatever you needed, was still there if you were willing to work for it.

Resisting the Bootstrapping Library

Library leaders called on a complex set of materials—architecture, computing infrastructure, specially selected workers—to bootstrap MLK. Every day, that library of the future was built around, and sometimes over or against, the needs of the present space in downtown DC, where one of the last few public spaces in the city—staffed by highly educated, helping professionals—served marginalized Washingtonians with few other places to go.

It is important to make clear the basic fact that though patrons vastly outnumbered staff, they lacked the organization, resources, or institutional support networks to dictate the library's design.[11] For this reason, and despite being the people putatively served by these reforms, patrons were not fully included as stakeholders in the planning and execution of the bootstrapping process. There was little incentive to include them. Patrons couldn't offer the credentials that iSchools did or the resources and legitimacy available from the municipal government—and their complex needs challenged the simple framing of the access doctrine.

Beyond what resources patrons did or did not bring into the library, the library itself also limited what patrons could change about the space, through choices made in its architecture, infrastructure, and the scripts for its technologies. The end result was increasingly stark divides built between the library of the future and the people it was supposed to empower. These divides could be quite literal, like the glass walls splitting the Dream Lab coworking spaces from the Digital Commons computer lab.

Still, resistance flourished. Patrons went beyond adapting to the bootstrapping library and created small alternatives to this professionalization of the space. The bootstrapping process could never fully succeed—indeed, as chapter 5 will show, it never can—and so the public space built into the library's ideological and architectural bones could always support fleeting moments of unprofessional place-making that largely did not engage with MLK's digital resources. This meant more than individual moments of watching porn that librarians couldn't decide what to do with; it meant creating collective spaces of play, collaboration, and rest that existed as an alternative to the professionalizing library—even if some were eventually subsumed by it.

First, there were plenty of places to play in the Digital Commons. Most important was a group of tables and chairs with no desktop PCs in the corner of the room, near the queue screens and the glass windows that looked out onto the lines forming outside Catholic Charities for its free meals. For 2013 and much of 2014, this corner was taken up, especially after school let out, by loud card games—mostly *Pokémon* and *Yu-Gi-Oh!* Friends, mostly young Black and Latinx men, met there every day and cheered each other on like at any other sporting event. Phones and computers were left to charge under the desks or against the walls, sidelined in favor of the main event.

But that is not "doing the library right." You wouldn't cheer on a *Yu-Gi-Oh!* match in your office. And so Jefferey, a mid-twenties colleague of Eugene's in the hipster contingent, replete with mohawk and mechanics coveralls, invited a friend of his who lived in a Maryland suburb to drive in on weekends and organize official *Pokémon* and *Yu-Gi-Oh!* leagues. The new Battle Subway Pokémon League was an official league of the Pokémon

Company International. It was advertised by the library, by players on Facebook, and by flyers showing cartoon monsters fighting in a DC Metro car. Robert, the organizer, brought official tournament jackets and badges to the Dream Lab one Saturday each month, when the startup employees weren't using the space. Space was also set aside in the foyer on some Thursday afternoons—though school was in session for part of that timeslot. With the raucous play contained to a themed space and time, librarians were free to crack down on gameplay in noisy corners during the week. The library's professionalism was secured.

Second, patrons built spaces of collaboration that differed from both the didactic, lecture-hall setup of the Digital Commons and the digital, professional collaborations of the Dream Lab, the Fab Lab, and more advanced classes. Collaboration was, of course, part of the plan for the Dream Lab's glass cubicles. Its largely White technologists were usually the loudest patrons in the room during the week, brainstorming on the cubicle whiteboards and video-chatting distributed work groups while the rest of the Digital Commons typed and whispered and had headphones offered to them by librarians if their music was audible.

Patrons acknowledged this segregation to me, but it hardly impacted their routine. Most walked right past it, to whatever desk they could find. But entrepreneurial collaboration still flourished outside of these approved spaces, though not of the digital, professional sort the library or the city government advertised—that is, what went on at Mapbox. For example, patron Ricky, who worked at a restaurant down the street, would roam the rows selling loose cigarettes. Sales of harder drugs, usually synthetic marijuana (aka K2) or crack cocaine, were frequent and more or less surreptitious depending on the client and seller. Library police were always looking for drug sales, taking a special interest whenever two people huddled together—even couples.

The library changed its own layout to crack down on these entrepreneurial collaborations. The sidewalk out front used to form a cozy corner with an alley separating MLK from the church next door, a secluded spot for sales or a nap along a busy block that got more crowded as lines formed for shelter shuttle pick-ups. This was a point of conflict for a neighborhood increasingly becoming a dining and residential destination and which already hosted a basketball stadium, a movie theater, and department stores among older office buildings. Residents of the ten-story Mather Studios condominium building across the street called a public meeting with the library, DC police, and their city council member in May 2014 to address what the meeting RSVP called "the degrading situation in front of the Library." That the meeting was prompted by complaints from the condo association, rather than long-standing struggles of library patrons, was again evidence of the uneven responsiveness of the library to different groups of stakeholders.

The report from that meeting gave residents guidance on what was (public urination, blocking private entryways) and was not (general loitering) illegal on DC sidewalks and provided instructions on when and how to alert police. That summer, MLK began leasing that cozy corner space to the Bike and Roll bicycle rental company. The company's bikes and storage units took over the space and invited more tourists into it. A conference on homelessness, with the Friends of the Library and local service providers— but no homeless individuals—followed soon after, as well as the hiring of a dedicated social worker, to some public fanfare (Sheir 2015).

Other forms of entrepreneurial collaboration in the library drew less attention from police and property owners. Oil men were a constant presence at MLK: Black men, often Muslim, with tiny vials of fragrant oils stored either in belts on their chest, jutting up against long beards, or in light wooden racks that can be carried under one arm. Pairs would meet up in a back row between rush-hour shifts, cataloging their products, planning pitches, poring over maps on their phones, working out sales routes. The library was their planning space, but, unlike the tech workers, they didn't use the Dream Lab's glass cubicles. Police mostly left them alone. Jackie, Mia's mother, drew on her many hours spent watching YouTube crochet tutorials to sell colorful knitted phone sleeves and wallets. Mia modeled the items to friends and directed business her mother's way.

There was also a robust culture of mutual aid. This included frequent trade in peripherals, as well as advice for speeding up used laptops or finding free software, collaborative efforts to combat a widespread condition of "dependable instability" in patron devices (Gonzales 2016). Friends exchanged for free; others kept themselves in cigarettes or bus passes by offering services like repairs or assistance loading songs onto MP3 players.

Mia acted as a one-woman community exchange from her favored spot in the back row of the Digital Commons, guiding friends to file-sharing services or through particular forms. One freezing day in January 2015, I entered the lab to find her sharing Pokémon tips and tricks with friends before moving to a more serious task: finding housing and figuring out whether her vouchers would continue. She and her friend Raquel, who was in the same boat, divided up the labor.

Mia had her own laptop and so took on the tasks that would've outlasted a session on the library computers: taking notes on DC housing policy and generating a list of shelters and transitional housing facilities for Raquel to call and ask about available beds. Raquel went down the list and reported the results to Mia, who recorded the data. Mia passed the contact sheet, now annotated with available beds, to other friends before turning back to Pokémon. Raquel shifted to calling temp agencies, taking notes on who answered the phone, their general demeanor, and the jobs available so that she could pass the information along to other library regulars. The library was a base for their collaborations, but, desks and Wi-Fi aside, actual library resources played little role in the process.

Beyond play and collaboration, the most important alternative use of the library space, especially for homeless patrons, was as a place to rest. This visible lack of productivity of course violated the professional script of the bootstrapping library, and so was constantly policed. For patrons, the Digital Commons was not only a place to apply to jobs or learn Excel. It was a place to check email between dishwasher shifts. Or a place to stop after a day program ends because most shelters kick residents out during the day. Or a place to sleep during DC's 100°F summers because neither shelter beds—"there's no such thing as relaxing at the House of Ruth," Mia told me—nor the sewer grates above the subway stop next to MLK are quiet, comfortable spaces at night and because many psychiatric medications are strong sedatives. Indeed, ambulance pick-ups from MLK were not uncommon, with patrons collapsing after an unseasonably cold night on the street or, in the case of Mia and Ebony's friend Josie, after being unable to secure new doses of seizure medication.

As with porn, librarians were conflicted over the issue of sleep. But the new focus on professionalism resolved the question for them: sleep not only was an unproductive use of the library's computer lab, but it discouraged others from productive activity. And so sleeping patrons were the most visible site of librarian discipline. Librarians patrolled, knocking on the desks of people dozing off with a loud "sir" or "ma'am," calling library police over if patrons did not respond. Elena explained how she worked through this conflict in her head:

> We are not allowed to sleep in the libraries. A library, whatever else it is, is "a place for lifelong learning"—that's kind of the buzz phrase of the moment. And it is and we want people who want to come in and use our collections to come in and use our collections and our resources and feel comfortable coming and using them.
>
> If all the tables are full of people with their head down asleep it's not super inviting for people who are there to use our services as more than just a building to sit in for the day. Which is why if you sit there with a book or a newspaper, you've kind of indicated you're interested in using our collections and our services but maybe you're tired [and so will not be woken up].

Elena could never decide whether library furniture counted as a public resource in the same way the computers and collections did. She quit DC Public Library right as the Digital Commons opened.

Elena's commitment to policing showed that patrons' sleep jeopardized the library's bootstrapping mission. Digital resources like the computer lab were what made MLK "more than just a building to sit in for the day." And their presence certainly brought people in. After all, when the Pharos system broke down in December 2014 and the computer lab's Dells were inaccessible, the library was open but empty. Patrons came to those machines to work or look for work, to manage social services, to connect and play, to compete or cooperate, to rest. Where else in the city could they find that sort of comfort and freedom, for free? But those tools and that space had to be used in a way that fit the bootstrapping library. Paradoxically, the library's openness pushed it further toward this more limited, professional vision. Librarians did not have the training or resources to meet all the needs patrons brought into the Digital Commons. And so they managed this local crisis of care by cracking down on forms of adaptation and resistance. Such discipline was not purely negative, cutting away at unacceptable uses of the space. Rather, in their day-to-day work, libarians' discipline expressed the acceptable uses of the space, reducing the diverse range of patrons' needs to a smaller, more tractable set that the institution was equipped to address This discipline was also forward-looking, producing the hope that professionalizing the library would secure a space in the future for the institution, its employees, and its patrons.

Conclusion: Partitioning Hope

The library bootstrapped in order to garner much-needed support from outside the organization and to make the complex social problems facing the organization more tractable for people inside it, remaking its operations, personnel, and space to support digital professionalism. The queue for computers made this strain and their management of it visible. Patrons needed to be, as librarian Grant put it, "using the technology to improve their lives." But resistance persisted. Patrons who needed something else from the library, and librarians who understood it as a truly public space, pushed back on the professionalization of the space. That resistance was largely overcome, as the issue of sleep made clear. MLK needed to be, as Elena put it, "more than just a building to sit in for the day."

There was a similar tension at play in the Du Bois charter school, the subject of the next chapter, where the stakes of student success were so high that the school, like InCrowd, operated in permanent beta, constantly reinventing itself in the face of new data. But MLK was literally rebuilt around this hope in a digital, professional future, shedding the old transactional space for a new transformative one. During this three-year $208 million renovation, MLK moved many of its services to other short-term homes. Planning for that process forced library staff to admit that the contemporary computer labs had failed and needed to be taken apart and put back together again. This made clear that bootstrapping does not have an end state; it is a constant process of organizational reform done in the hope of finally finding the right mix of digital tools and skills that will bring the people on the margins of the information economy to its center. As the crackdown on rest spaces showed, preserving this mission, so important to the future of the organization, could mean refusing the needs of those same marginalized people.

Grant, a Black librarian in his late thirties, described this process to me. He wore a suit to work, in contrast to the hipster contingent, and patrolled the Digital Commons with authority. He was resigned to the renovation, maintaining that Martin Luther King Jr.'s name was disgraced by the building. It should have been torn down and rethought a long time ago, he said, and would have been if it were not for its landmark designation. He believed patrons were not being served by the library, or any DC government agency: "Those patrons need help but we're not in a position to help them at all. As a matter of fact, I feel like a lot of our staff here feel we need to entertain them."

Looking back through the Dream Lab's glass walls one afternoon, he walked me through a game he often played—not dissimilar to April's gold stars and paint buckets. Grant estimated that only four or five of the seventy-five users he walked past to get here were working on job applications or resumes, with the rest playing video games or on social media. "Been doing that all morning. That's what they were doing yesterday, that's what they're going to be doing tomorrow." He sighed and nodded toward Shawn, watching *Dragon Ball Z* cartoons on YouTube. They had a long-standing relationship. Grant had gotten Shawn some clothes and money, acted as a reference, and helped Shawn map out library classes that would refine his artistic skills. "When he sees that you're trying to do something with yourself, like fill out resumes in there or go to school, those librarians, they'll be there for you," Shawn said. It was clear to him that Grant cared and that his care went beyond the narrow frame offered by the bootstrapping library.

Shawn looked up to Grant and appreciated the librarian's commitment to service, but he never took those classes. Life got in the way; being poor in the city is a full-time job filled with endless appointments, queues, and moves. Grant suspected Shawn's depression didn't help matters. Ebony had returned to school to get her GED, showing off her report card to her crew in the Digital Commons each quarter. Grant admired her for that, but he was not angry at Shawn for not following Ebony's example. Firms like InCrowd that were changing DC, after all, produced culture as much as they produced software, and Grant recognized that a homeless Black man with no college and scant work history would have trouble being accepted into that culture—even if he did master Adobe InDesign.

Grant observed a conflict between what the digital resources were meant for—professionalization—and what they were usually used for—rest, play, and collaboration. He taught classes with the 3-D printers or the Espresso Book Machine and felt that "there is innovation happening" in that classroom or the offices next door. But he knew that because of the "stigma associated with this building," the younger, wealthier, Whiter crowd would "make a beeline straight for what they have to do and they don't hang out." The present space was not for them, but the future one might be. This was true even for the Friends of the Library. When they visit, "We're told to tidy up and keep everyone quiet, so they don't scare the good White folks. And I think that's absolutely disgusting. And I resent it."

The hope had been that professionalizing the space would empower regular patrons—but those resources and that mission were best embodied by the sort of White professionals who irregularly visited the library. The solution, Grant said, would be to further partition the space: "Certainly, moving forward with the renovation, everyone seems to be pretty clear that this did not work. The crowd they want to have—this crowd, our everyday patrons, are not using the services we hoped they would use. So, if they want to use the computer all day, mess around, that's fine, it's just going to be in a different space. And those that want to get serious with technology and support technology, there will be a dedicated space for that."

What this meant in practice was that during the renovation, some of MLK's computers, training programs, adult literacy classes, and collections would be distributed across existing branches and temporary Library Express locations to serve the homeless patrons dislocated by MLK's closure. It would be a different story for the places where "innovation was happening." The Fab Lab, for example, had special electrical and ventilation needs that couldn't be served by a branch, so it hit the road as an ambassador from the library's future. It stopped in the DC government municipal building on the bustling Fourteenth Street NW corridor, and then in a shipping container in the gentrifying NoMa[12] neighborhood. The container was sponsored by the NoMa Business Improvement District, the Institute of Museum and Library Services, the DC Public Library Foundation, and Uber.

Plans for the new MLK retained some elements of these temporary partitions. There would be no dedicated Dream Lab in the new Digital Commons; glass cubicles would be scattered throughout. The new library would allow for more crowds and noise on that ground floor, and then increasingly, in Grant's words, "get serious with technology" on higher, quieter floors. The Fab Lab and a new Creative Lab took pride of place in the heart of the rebuilt A Level. Dedicated collaboration and coworking spaces would be moved to the third floor between executive and staff offices. In Mecanoo Architecten and Martinez + Johnson's winning architectural renderings (2014), the before picture of the Popular Services collection on the first floor showed a dim, frontal shot of three older men at rest, full bags at their side. The after picture showed a collection of young, multiracial families and professionals striding through the bright, open space and around its café. Grant and his colleagues suspected that in practice this meant there would be spaces for daily patrons like Shawn and Mia, and other, separate spaces for more professional visitors. This is the bootstrapping library. It is not just a future transformation of physical space but an active reorganization in the present of people, technology, and ideas. Thus we see that preserving the hope in personal computing to change the future requires partitioning it from today's messy needs. The homeless patrons watching YouTube or dozing off in the back of the Digital Commons did not fit the hope of the Dream Lab or the Fab Lab upstairs. At the time of this writing, the library was still being built, both by construction workers and by the people who will eventually fill the space. But for now, and, if the renderings are to be believed, in the future, those rest places, collaboration places, and play places that patrons built will be physically segregated from the startup workspaces, the seminar spaces, and the transformative technologies that will form the heart of the new library.

4   Flexible Classrooms

The teachers who worked with the senior class at W. E. B. Du Bois Public Charter High School met once a week to strategize. Everyone's laptops came out, collaborating on notes hosted in the school's cloud-based Google Apps for Education system. Every month or so, this expanded into an all-staff meeting that also drew in Principal Carroll, other high-level administrators, college counselors, and technology specialists. They usually met before the school day began, in the Think Tank.

The windows taking up one side of this large, open room looked out onto the school's field and its northeastern DC residential neighborhood. About eighty feet long and thirty feet wide, the front half of the Think Tank was an instructional space with furniture and fixtures designed to be moved around for different classes' needs. The back half was an open workspace with cubicles, but also beanbag chairs, outlets, and study nooks where seniors could work on their own schedule, tucked away with one of their school-issued laptops. It looked like InCrowd's office or MLK's Dream Lab.

In December 2014, the senior teaching team was worried. A significant minority of the first senior class was not on course to graduate. Only sixty strong, those seniors were prototypes for the school's methods, and their success or failure would be a referendum on the school. Du Bois's mission was to get every student into "the college of their choice." That mission was operationalized through seniors' graduation and admission rates and by other grades' standardized test scores and collective GPAs. Those metrics would prove the school's success or failure. The mission was supported by an intensive, personalized curriculum that made heavy use of information technology: a laptop for every student; experiments in making video games or robots; and a robust digital infrastructure called SchoolForce, which tracked grades, attendance, and more, shared the information between teachers, and pushed alerts to administrators and parents. Du Bois and other DC charters had agreed to beta test SchoolForce for Acumen Solutions[1] as part of a federal grant. It was built from the API for the popular Salesforce customer relationship management software that was used by InCrowd.

But technology wasn't enough. The school also prided itself on a particular set of racial justice values that supported its mission: an emphasis on racial diversity in curriculum planning, a restorative approach to discipline that avoided suspensions and expulsions, and a general push for the largely White faculty to focus not on solving perceived deficits in their Black and Latinx students, but on empowering them to develop their strengths.[2] The different pieces were supposed to work together: technology empowered disadvantaged students to train for success in a hostile world.

School technology and school values were chosen through internal deliberations among teachers and administrators. But the success or failure of Du Bois's mission—those graduation rates, college admissions, and test scores—was assessed externally by the leadership team, by the DC Public Charter School Board (DCPCSB), and by the philanthropies supporting the school, all of which wanted to see a return on their investment into students' human capital. This chapter explores how these governance dynamics influence the bootstrapping process, manifesting in simultaneous, intersecting conflicts over the school's technology and values.

Teachers tried to build a high-performance academic culture, whether by modeling technology use themselves or encouraging constant connectivity in their students. But they stumbled, for reasons largely beyond their control. Failure was not an option, so more direct modes of discipline were eventually pursued through SchoolForce. Ultimately, the school's digital infrastructure was more responsive to administrators than teachers or students. It was used to subvert the school's values to its mission. In seeking to fulfill that mission by any means necessary, the school ended up marginalizing many of the students it sought to serve. The access doctrine was meant to serve students, but it necessarily had to serve the school first.

It was easy to see these conflicts in the December senior team meeting. When I walked in, teachers were debating a question on the whiteboard: "Why don't we have a high-performance academic culture?" They were worried. They had spent years trying to encourage their students to take up good academic habits. Teachers were supported in their efforts by the resources that came with one of the highest per-pupil budgets in the city (almost double that of a typical DC public high school, driven by corporate partnerships, donations, and grants won by a full-time fundraiser). But they weren't seeing the results they wanted. In that meeting, phones, as usual, were the locus of every faculty complaint, the site at which good students were demarcated from those with "bad academic habits."

Lead college counselor Byron, who was Black, reprimanded White calculus instructor Ryan for questioning whether teachers' expectations were skewed because they all went to "top tier colleges."[3] Byron insisted, and the faculty largely agreed, that "we should not be referencing our college experiences" and that a high-performance academic culture—setting and completing goals independently, using technology deliberately, managing multiple demands on one's time, pushing peers to focus on the same—would serve students just as well at the University of the District of Columbia as at Harvard. That some students might not choose to attend college or might not be able to afford it despite counselors' best efforts went unsaid.

Faculty decided they would begin encouraging a high-performance academic culture through an intensive data-tracking regime based in School-Force. Students would be scored on habits like checking their phone. Teachers would push them to track their scores, change their behavior accordingly, and encourage their friends to do the same.

These proposals were insufficient, or at least they did not assuage administrators' fears. In January 2015, as the first semester was drawing to a close, an all-staff meeting was called. Everyone met in the Think Tank before first period. The largely White faculty asked the largely Black and Latinx students[4] who showed up early to do their homework to please move downstairs. Principal Carroll reiterated the concern: if graduation were today, almost half of the senior class would not have the GPAs necessary to receive their diplomas.

The stakes had been raised. The school's leadership team—composed of the Board of Trustees; the head of school; middle, elementary, and high school principals; and other administrators in fields like instructional design and fundraising—had reached into the SchoolForce backend, seen the spread of GPAs, and officially raised the alarm. Teachers had been pointing this out for months. Now they complained that they didn't know who these board members were and that they didn't understand why students on the ground were struggling to live up to Du Bois's high standards. Those high standards included enrolling all students in Advanced Placement (AP) classes; providing individualized college application guidance; and training for both national exams (e.g., SATs, AP exams) and, for younger students, local high-stakes tests: the DC-CAS and, beginning in the 2014-2015 academic year, the Partnership for the Assessment of Readiness for College and Careers (PARCC) subject area tests. The leadership team's alarm was exacerbated by year-over-year dips in SAT practice scores and the DC-CAS scores on which every DC school had been judged.

Du Bois was in trouble. The numbers were off. Technology had become a distraction, rather than a tool for academic success. Teachers' resistance to the leadership team prompted a debate over Du Bois's methods and values. This included the usual negative comparisons, from both sides, with a DC public school system that most staff agreed could not provide its students the innovations they needed. They saw a more hopeful future in this building—itself a shuttered public middle school, bought by the leadership team years back, renovated, grown out, its walls inside painted with the school's colors and inspirational slogans.

Principal Carroll promised to bring faculty concerns to the leadership team. But they were all called into a meeting with the board in the gym the next day anyway, to repeat the message that the numbers were off and change was needed. The leadership team wanted teachers to double down on the changes they'd proposed to encourage a high-performance academic culture. SchoolForce surveillance would intensify. This individualized tracking seemed to embrace the sort of deficit thinking the school's values opposed. Laptop and phone habits would be a major focal point of disciplinary actions in senior classes, which seemed to run counter to the school's restorative justice principles.

The weight of first-quarter grades would be retroactively lowered. The last day of the second quarter would include several hours of credit recovery. Principal Carroll said that day "was about forgiveness, not justice" and that it gave students precisely targeted opportunities to revise certain assignments or retake certain tests to boost their individual GPAs and get the collective up to a level that would satisfy the leadership team. Some teachers boosted grades regardless of whether the revised assignment was actually completed to a higher standard, or just added a few points to students' overall grades.[5] The mission was that important.

Although these experiments were carried out under duress, they fit a general pattern of a school that prided itself on innovation. The work of Du

Bois's teachers and administrators was marked by constant experimentation: experiments in implementing the one-to-one laptop program, experiments in the content of their homeroom period, and even experiments in the number of seconds available to transition between classes. New pedagogical techniques were rapidly prototyped, and kept if they succeeded or discarded if they failed. Because the stakes for their students were so high, every piece of technology and every piece of the school's values were subject to revision in service of its mission. New curricula and technologies were constantly tested, tried, and embraced or rejected. It was a school in permanent beta.

But these frantic winter meetings showed that the power to start or finish an experiment, or declare it successful or unsuccessful, was not evenly distributed throughout the school's stakeholders. Ultimately, the digital infrastructure—SchoolForce—that supported Du Bois's experimentation was under the control of these administrators, not staff and students. That infrastructure was used first to revise anything that did not support the school's local version of the access doctrine and second to directly discipline students' academic habits. Similar to MLK, the panoply of social problems faced by Du Bois's students—hunger, homelessness, police and neighborhood violence, disability, segregated labor markets, gentrification—were narrowed into the more tractable problem of academic performance. For individual teachers, this meant disciplining students' phone use. For the senior team as a whole, this meant a push for higher grades and test scores.

Where MLK became a bootstrapping library over time, Du Bois was born a bootstrapping school. Unlike public libraries, around for more than a century, charters were created for an era of extreme economic uncertainty. Like the startups with which they share an aesthetic, charters' identities are built on those uncertain conditions and their attempt to master them through technological skill—and to help others do the same. But schools can't pivot like startups. Du Bois couldn't remake itself to the degree InCrowd could.

Understanding these dynamics requires first some background on charter schools and their embrace of the access doctrine. I then return to the everyday life of Du Bois's first senior class. Teachers first experimented with a series of implicit measures to build a high-performance academic culture: modeling correct technology use themselves and encouraging constant connectivity on their students' part. The leadership team's alarm made clear these efforts were insufficient. So teachers turned to SchoolForce, an infrastructure ultimately beyond their control, to provide the discipline necessary for a high-performance academic culture. When tensions arose between the school's racial justice values and its mission to upgrade students' human capital, the SchoolForce digital infrastructure favored the latter over the former. It had to. The problem of human capital deficits in the information economy was built into the school. As in MLK, the access doctrine had to be preserved for the sake of the organization's survival, even if that meant neglecting the needs of the people the organization was built to serve.

A "Beacon of Hope"

Each charter school is its own experiment in a particular educational philosophy. As public-private partnerships run in parallel to the DC Public Schools (DCPS) system, they receive city funding equivalent to that of traditional public schools, but generally secure additional funding from outside school-management networks (e.g., KIPP, Rocketship, Success Academy), as well as competitive grants secured by dedicated fundraisers. Admission is theoretically open to any student who wins it through the lottery system (Chandler 2015a). While I was at Du Bois, charter schools educated about 44 percent of DC students, with most of the rest attending traditional public schools in their neighborhood (DCPCSB 2020).

Du Bois prided itself on being the best the charter school movement had to offer. It avoided the exclusionary habits that put some of its peers in the news: English language learners and special education students were always admitted to Du Bois, and they counted toward the test scores against which the school was judged, in contrast to some other schools where the themed curricula could not or would not support these students (Chandler 2015a; for national context, see also Scott 2012; Simon 2013). It hired no new teachers straight out of college. Du Bois also used a variant of the Reconceptualizing Early Childhood Education (RECE) framework to inform its disciplinary policy. This meant there were no disciplinary suspensions and expulsions (except in the case of violence); instead teachers used redirections meant to empower disruptive students to make behavior changes and recovery time to help them work through it. This stood in contrast to the "no excuses" model of some prominent charters, which produces high rates of suspension and expulsion and exacting standards for how students sit, stand, and make eye contact with their teachers (e.g., Golan 2015; Green 2016; Lack 2009).

But Du Bois was still firmly within the charter school movement, insofar as it attempted to boost the human capital of urban, working-class Black and Latinx youth through an entrepreneurial approach to education that linked classroom experimentation, digital literacy, STEM knowledge, and data-driven accountability. Charter schools are hallmark institutions of the access doctrine. Charter advocates not only position education as the intervention point for populations left out of the information economy's riches, they also position public schools themselves as broken, industrial-era institutions that need to be creatively remodeled along the best practices of private industry, especially the agile, creative practices of the tech sector. Public education, charter advocates argue, needs a pivot.

Charter activism is a key part of the neoliberal rethinking of welfare state institutions, wherein they are no longer universal social democratic benefits for all citizens or bulwarks against periodic crises, but sites where the human capital of the citizenry can be enhanced to a level sufficient for competition in the information economy. For example, the Center for Education Reform's *Mandate for Change*, a 2009 lobbying document addressing the incoming Obama administration, prominently featured Kevin Chavous, former DC City Council member and founder of Democrats for Education Reform, advocating for charters as a "beacon of hope for parents and students alike" (20). Chavous continued, "We must leave the Industrial Revolution behind and embrace a new model of public education. One single approach no longer works with all children" (23).

The success of this project is typically measured through standardized test scores. Municipal bodies that grant a school's charter—in DC, this is the DCPCSB—do so for an array of school operators, the bids of which are based on their unique capacity to raise these scores and ensure student success. Charters thus compete with each other and traditional public schools for students, funding, and political approval. Unsuccessful models, the theory goes, will be shut down, while successful ones will be replicated.

There is little evidence that charters outperform traditional public schools on these standardized tests.[6] However, there is substantial evidence that they suspend and expel Black and disabled students at a higher rate.[7] As Chavous made clear in the mandate, charters in urban school districts are intended to replace what education reform activists see as a broken public schooling system that rewards incompetent teachers with lifelong (unionized) jobs while sacrificing the educational (and thus economic) futures of vulnerable working-class students of color.[8]

DC was ground zero for the education reform movement generally and charter schools in particular.[9] That distinction was earned in the years following the more general crises of underinvestment and depopulation that Anthony Williams managed as mayor. Williams was succeeded in office by Adrian Fenty. Fenty's tenure was the start of a major growth period for the city, which, owing largely to the presence of the federal government, survived the recession better than most, seeing a boom in population and the eventual loss of Chocolate City's Black majority.

Seeking solutions to structural poverty and the "achievement gap" between DC public schools and schools in its richer, Whiter suburbs, Fenty installed education reform veteran Michelle Rhee[10] as the schools chancellor in 2007, after removing the elected school board's formal governing powers. A popular documentary promoting charters—*Waiting for "Superman"*—highlighted Rhee's reforms. President Obama fêted the children from the film at the White House.

Because of Congress's special regulatory and legislative powers over DC, the city's public schools have been a constant site of federal experimentation. Rhee, wielding a broom on the cover of *Time* in 2008, ready to clean up DC's schools, reversed this trend and pioneered city-wide changes that added up to "the most aggressive reform effort of any unified urban district in America" (Osborne 2015, ii). DC become the model for education reformers across the country. Rhee closed dozens of schools with low enrollment, initiated a teacher-evaluation system linked to student test scores, fired hundreds of teachers based on that system, used a new tenure contract with the teacher's union to incentivize millions of dollars in private philanthropic donations (many, like those from the Gates Foundation, for educational technology procurements), and approved dozens of new charters. Many, like Du Bois, took over shuttered public school buildings (Craig and Turque 2010; Leman 2013).

The mass firings of mainly Black teachers and the public fights with the Washington Teachers Union generated mass disapproval for Rhee from black Washingtonians (Turque and Cohen 2010). There were deep, historical reasons for this. Until 1977, the District lacked a comprehensive university. For decades prior, DC's teachers' colleges provided the most direct route to professionalization for residents, and so an attack on teachers was seen as an attack on the Black middle class (Schwartzman and Jenkins 2010; Tavernise 2011). The 2010 election became a referendum on Rhee's tenure, and Fenty was voted out. Rhee left too, but new mayor Vincent Gray retained and promoted her second-in-command, Kaya Henderson, who had previously worked with Rhee at the New Teacher Project and was largely responsible for developing DC's teacher-assessment system and negotiating the 2010 union contract. Charters are not included in this contract and were, until late 2019, entirely nonunion (Craig and Turque 2010).

DC's charters differentiate themselves through various themed curricula. These might be experiments in language immersion, arts education, and leadership training—but most often in college preparation, STEM education, and innovative deployments of educational technology. Students win entrance to a particular charter through participation in an annual lottery in which their family ranks a list of choice schools. Participation is open to any family with the time and resources to research schools and fill out the online application.

Du Bois rejected parts of this political legacy through policies like restorative justice and the hiring of veteran teachers. Still, everything the school did, down to mid-year adjustments to the seconds available for transition time between classes, was meant to build an engine of social mobility for Black and Latinx students. That mission demanded a lot from teachers and students. "Du Bois is super academic. It's almost rigidly academic. It's a place that is designed to make kids successful in the systems that we have in the world right now," teacher Amanda told me. Senior Irene said the "main idea" of Du Bois was "pushing you until you can't hold it in no more, like you literally have to explode because of this insane workload, right?"

"College of your choice" was the goal for every student. The personalized learning meant to get students to their college of choice was facilitated by the, novel at the time, one-to-one laptop program and an intensive data-tracking regime. Constant experimentation was the norm. Test scores, and to a lesser extent GPAs and graduation rates, remained the make-or-break measure, the thing that prompted bootstrapping. The rigidity Amanda saw was certainly present in their mission and its standards, but the school was, above all else, flexible in meeting those goals, which demanded much from teachers like Amanda and students like Irene.

Flexibility was key to Du Bois's project and to the charter movement more generally. Like startups, new data prompted new experiments in the shape, strategies, and structure of the organization. This uncertainty about what the future held for the school was embraced in the name of preparing marginalized students for a general environment of economic uncertainty that they would face upon graduation. They would do whatever was necessary, arguing that traditional public schools could not do the same. When Clinton and Gore suggested in 1993 that "schools can themselves become high-performance workplaces" to train the digital professionals of the future, something like Du Bois was probably what they had in mind.

I entered Du Bois suspicious of charters. I found a great school. But the access doctrine motivates charter schools to an even greater degree than it does public libraries, and thus everything great about Du Bois was subject to revision, depending on what the data said. As in MLK, the problem of poverty became a problem of technology—even as every teacher and student knew the problem was far more complex than what skills or tools students lacked. And while the changes to the school's structure were perhaps not as stark as MLK's full-scale renovation, they were far more frequent. While circumstances demanded they both bootstrap, Du Bois had a license to experiment that MLK lacked, supported by fine-grained analyses of student and teacher data that had no corollary in the public library. Du Bois's racial justice values were one experiment, something that could be discarded if it wasn't enhancing students' human capital.

The refrain that education reform is "the civil rights movement of today" is a common one among charter advocates (Scott 2011). Even if they didn't always say it in those terms, Du Bois staff certainly acted as though everything was on the line and that, for their students' sake, they needed to act with the fierce urgency of now. This prompted constant experimentation. Sam, a senior team leader, likened the process to building a plane while flying it, with summer the only time when teachers and administrators could think in the long term. Irene said that by the end of senior year, she had seen three iterations of the one-to-one laptop program and many other experiments besides. She was tired of being experimented on: "The first class, the students are like little lab rats."

Nowhere was this experimentation more obvious than in the repeated conflict over student technology use. Staff wanted students to develop a holistic view of their academic performance, connecting different classes with home life, school life, and their future academic and professional goals. The laptop was the tool to do this. The phone was the tool preventing this;

it made the wrong kinds of connections. Informed by SchoolForce, and under pressure from the leadership team, staff experimented constantly with new ways to encourage a high-performance academic culture through technology use. Although students were the objects of experimentation, both of these modes of discipline—encouraging laptops and discouraging phones—were also stressful for the teachers who had to carry out the ever-changing tactics deployed in the name of Du Bois's mission.

Throughout the year, teachers experimented with two different ways of encouraging a high-performance academic culture: by modeling it in their own behavior and by encouraging it through constant connectivity. The leadership team's January announcement made clear that these were insufficient. So in the second semester, teachers increasingly used the school's digital infrastructure—chiefly, the SchoolForce student data system—to provide more explicit discipline. Both role modeling and digital discipline provoked a direct confrontation with the racial justice values at the core of the school's identity. There was much room for debate, except with School-Force, which was ultimately out of the control of any individual student or teacher.

Technology for a High-Performance Academic Culture

Like MLK's librarians, Du Bois's teachers worked hard to define and enforce the correct way to use a PC as part of their efforts to build a high-performance academic culture. This pedagogy appeared in two ways. First, implicitly and throughout the year, teachers modeled a high-performance academic culture through their own technology use and encouraged a culture of constant connectivity in their students—similar to InCrowd's 24-7 workstyle. These experiments in a hidden curriculum were largely under teachers' control and they are the focus of this section. Second, it appeared through explicit data-driven discipline that used SchoolForce to "penalize phone use other bad academic habits."

Data-driven discipline picked up pace after the leadership team's January ultimatum and, because this infrastructure crossed the whole school, it was ultimately under the leadership team's control. This is the focus of the next section. As in MLK, these efforts to discipline personal computing prompted a debate on the nature and purpose of the institution: What was schooling supposed to accomplish, for whom, and how?

The first class of seniors had been through three distinct implementations of Du Bois's one-to-one laptop program: from personal Dell netbooks
to personal Acer Chromebooks to laptops in carts that rotated through
classrooms. In ninth grade, students had received Dell 2120 netbooks.[11]
They put down a fifty-dollar deposit and signed a rental agreement dictat-
ing the rules for responsible use during the school year.

Dell stopped producing 2120s, so the following year, ninth graders were
issued Acer Chromebooks, optimized for the Google software most classes
already used. Sophomores who had lost their laptops the previous year were
issued Chromebooks. Teachers reported having no training or consultation
for the new machines. By the time I entered Du Bois, most seniors no longer had their Dells; they'd been lost, stolen, or just broken. Chromebooks
still circulated, but at least half the seniors lacked any computer. Those few
(five or six) seniors with the family resources to do so brought fully featured
laptops or iPads from home.

The school replaced the computer the first time it was lost, stolen, or
destroyed, but required a $500 replacement fee the second time. This is
what happened with senior Martin, who spent most of his out-of-school
hours working at Chipotle or using the camera he had bought with Chipotle wages. He had his netbook for two years, until a friend sat on it and shattered the screen. Many families were unable to come up with $500 on the
spot, and some struggled for the initial fifty dollars. Other computers were
simply lost on the bus or at the library, stolen, or, enterprising students told
me, sold once their serial numbers were scraped off.

Principal Carroll told me Du Bois had absorbed a nearly $35,000 loss on
school computers in the 2013–2014 school year. Du Bois was small, but even
a traditional public school of similar size would have struggled to bear that
loss. Du Bois's grants, donations, and bond sales helped it to survive.[12] Still,
Principal Carroll and other administrators knew that change was needed.
As part of Du Bois's empowering approach to education, it had not planned
for theft or loss. It had largely let students, many of whom lacked a clear
workspace at home, do with their computers as they pleased. There was
no equivalent to the Intro to PC Basics class Betsy taught at MLK, which
would have helped orient students to their machine and teach them to care
for it. There was no time for a subject that would not contribute to local or
national standardized tests. By the time I entered Du Bois in the fall of 2014,
school computers largely lived in mobile carts. Some seniors retained their

older machines, but ninth graders were not issued their own—a relationship that, as we will see, paralleled shifts in Du Bois's culture and curricula.

Hidden Curricula

Used correctly, school laptops could be the centerpiece of a high-performance academic culture. Students understood that laptops were meant to implement school values in a way that supported the school's mission. Senior Daniella told me that what made Du Bois Du Bois was "the fact that they use technology a lot" to support a "mission" where "it doesn't matter where you come from, your background, or your socioeconomic status, or your home life—you all go to college." Compared to the private religious school she'd attended in ninth and tenth grade, Du Bois was "a really open-minded school. They make their planning around technology and that's different." She was inspired by the mission and committed to the school's values.

But defining correct use was difficult, and enforcing correct use was harder still. And every small struggle over laptop and phone use became a small debate over the purpose of schooling and the place of technology in it. Teachers complained about students carrying their laptops by one corner, tossing them into a Think Tank beanbag chair while they made the rounds to chat up friends. But students were treating computers like they treated their phones. Shattered phone screens were a common sight—but these were interpreted by students as accidents or personal failures. Laptop failures were instead interpreted as a failure of the institution itself. That an app would not load or that the case easily chipped was, for students, an indictment of Du Bois. As with her regular complaint that the school lacked a dedicated art class, Irene said her slow laptop signaled that Du Bois wasn't a "real school" and could never live up to it stated values.

Teachers were not exactly careful with their laptops either, although that was more out of the necessity forced by a day scheduled down to the minute. Like tech workers, they were constantly shifting tasks and changing their approach to those tasks based on incoming data. Unlike tech workers, they were not judged on their own performance, but on that of their students. Unsurprisingly, an enormous amount of stress resulted from this position, wherein an urgent mission was realized by a group of teenagers for whom you cared but could not ultimately control. Teachers' own work habits helped to both manage this stress and provide an implicit role model of appropriate computer use.

Teachers walked between classes with their laptops in one hand, screen open, coffee in the other hand, placing them down on a desk when they entered a room so they could immediately start up a PowerPoint presentation, explore a student's grades with them, or shoot off an email or two as students completed a warm-up activity. Opening the laptop back up and logging back in would take precious seconds, seconds that administrators counted down as students transitioned between classes or between activities within classes. These faculty performances showed that professionals had their work machines glued to their hips and ready to go and that there was always a way to slip a microunit of work—an email, a grade change, a slide adjustment—into any time that did open up.

High-achieving students picked up on this role modeling. Irene and her friends Rochelle and Liu regularly worked over lunch, netbooks open, headphones on, checking SchoolForce or the progress of college applications. Lunches sat next to the laptops, at their elbows, either the school lunch or some chicken wings from a corner take-out spot; off-campus lunch was a privilege granted based on grade level. They embodied a high-performance academic culture: working independently, using technology to reach goals they set themselves, juggling competing demands, and encouraging their peers to do the same.

Education scholars call these sorts of informal messages about personal conduct, delivered outside the normal syllabus, a *hidden curriculum* (Anyon 1980; Calarco 2018; Giroux and Penna 1979). The hidden curriculum shapes attitudes and behaviors for the working world outside school, teaching students everything from how to sit to how to ask for help. Students were aware of the racial and class implications of the hidden curriculum. Irene described it this way: "They're so good at trying to prep us to be like preppy White folk....My thing is, like, we're in America, right? And I'm Hispanic, so the ideal would be that everyone—it doesn't even matter what race you are—we all have to be like the White people because apparently it's assumed that all White people are successful and whatnot. To get far in this country, you have to be like them, if you want to be up there, you have to be like them. I think for me, it's kind of, like, irritating, because that just takes the fun away and everything." Irene recognized that the school had the power to impose these hegemonic standards, where a high-performance academic culture was equated with whiteness and professionalization. Her teachers needn't speak this message for her to hear it.

Like many of her peers, Irene absorbed the hidden curriculum and took pieces of it for her own needs. She worked independently on her computer during lunch and free period, whether typing up a story for a school newspaper or making art for her sister's quinceañera. But she had also learned email etiquette and used it to contact teachers and request greater representation of Latinx people in the school's curriculum. And she and everyone else knew the posture—back straight, headphones on, hands on keyboard—that their teachers associated with productivity. Keep that posture and you had a good shot at spending free period on video games or Twitter without teachers noticing, compared to if you were slouched in a beanbag chair in the corner of the Think Tank.

Teachers did not need to make the link between whiteness and professionalism explicit for students. It was just what happened when White professionals sought to use their own workstyles as models for Black and Latinx students. In fact, teacher Clara's sociology class studied this same process in its unit on inequality and education. However, Black and Latinx students resisted, especially in those moments when the hidden curriculum went beyond role modeling and into direct regulation of the school space.[13]

The Think Tank was an important site of cultural instruction. Over time, teachers restructured the space to better support the school's mission. As in MLK, a desire to professionalize personal computing led to restrictions on who could use the space and how. Initially, the cubicles and beanbag chairs in the back of the Think Tank were meant to be independent workspaces for seniors on their free periods, while classes continued in the front.

But teenagers were teenagers, and they turned the workspace into a social space. Ryan and Sam would regularly pause their calculus or physics instruction in the front of the room to walk to the cubicles and beanbag chairs in the back and quiet students playing Flash games or YouTube music videos on their free periods, often telling them that they were not making the best use of their own time or supporting their peers who were trying to apply themselves. Principal Carroll checked in on the Think Tank several times a day, with greater frequency as the year went on. Walkie-talkie in hand, she would give a nudge to a student who was napping, or separate a couple petting each other in the corner and then watch as they took out their homework packets, plugged their headphones into their laptops, and got to work.

This intensified over the course of Du Bois's academic year, particularly after the leadership team's January ultimatum. Teachers and administrators only allowed those students who they thought were going to focus on homework into the Think Tank for study hall after school. Students were not creating workspaces independently, so teachers and administrators stepped in and did it for them. This is not to say that students were not putting the Think Tank to good use—study hall was a safe, supervised, after-school space, even if you weren't studying—but the bootstrapping project demanded one particular, professional use of it. Beginning in the second semester, school security guards began enforcing correct use of the space, kicking out students who weren't doing homework.

As in MLK, there were limits on this space-making project. At Du Bois, an older public service culture was literally built into the school, placing hard architectural limits on what bootstrapping could accomplish. There was no Think Tank in the traditional public school that had previously occupied the space. It had not been built as a coworking space, and so there were never enough outlets for students who needed to plug in their computers. And the ones that were there were often in locations that did not facilitate the sort of work Irene, Rochelle, and Liu did: sitting down to focus on the laptop. Most outlets were along the walls, naturally drawing students to them to charge their phones and lay in the sun. There were no outlets in the center of the Think Tank, where students collaborated on classwork or were directed to cubicles to focus on their own work—and it was a little too far for extension cords. So when Principal Carroll or anyone else directed a student to sit down and focus on their laptop, there was a good chance they couldn't comply because their computer was not charged.

Other limits emerged from students who, like library patrons, carved out their own places in the Think Tank. Irene had a place to work on her paintings, for example. Savvy students like Martin figured out how to breach the firewall and play video games with friends when it appeared they were working: using proxy servers, slight changes to URLs, or the creation of new user profiles on the Chromebooks that had teachers' internet privileges. In response, and as in MLK, teachers and administrators patrolled the Think Tank and directed students away from these distractions and toward schoolwork.

Students also built places in the corners of the Think Tank to buy and sell drugs, mostly pot. It usually went unnoticed, but the walls of that safe place could quickly collapse. A rumor was started that Martin was selling, so Principal Carroll, one of her assistant principals, and school security took him out to search him and his locker. They found nothing. Afterward, they told him he could have exercised his right to object to the search.

Some of these student efforts were stymied by the school, but Du Bois's values supported others. Clara supported her sociology students as they staged a walkout that began in the Think Tank, in protest of Darren Wilson's acquittal for the murder of Mike Brown. She also chaperoned students as they went to Black Lives Matter demonstrations downtown. But in the spring, the Think Tank was frequently taken away from seniors to set up workstations for tenth graders taking the PARCC standardized test or for those students taking AP tests. Clara's sociology class had critiqued these sorts of tests as tools for racial stratification. They might run against Du Bois's values, but they were essential to the school's mission.

In support of that mission, every single laptop in every single cart was requisitioned for the month of March so that they could be secured, set up for the PARCC exam, and then issued to a tenth grader to take the test. Teachers only found out two weeks ahead of time. The Think Tank became a testing ground. It stopped being a collaborative workspace, much less a social space, for the duration. Desks were separated out, and students sat down and surrounded themselves with school-issued cardboard walls that kept them from peeking at each other's screens. Irene had to borrow a counselor's computer to finish scholarship applications because hers had been confiscated for testing. The hidden curriculum became much less hidden at that point. Like any school, social life in Du Bois was rich and varied. But the Think Tank's transformation made clear that when the chips were down, mission mattered most, and the horizons of school life were radically narrowed to fit that mission. These social limits appeared as limits on school technology and school space.

Teaching Presence Bleed

Earlier in her career, Melissa had taught at a Houston charter school. There, she was coached to refer to students, teachers, and administrators collectively as "the family." Students were given her cell phone number, and she was expected to return evening calls requesting homework assistance. To incentivize teachers, bonuses were allocated based on student scores on Texas's high-stakes standardized tests. The cell phone requirement in Houston was an example of what Gregg (2013) calls *presence bleed*: the extension of the workspace into nonwork spaces through information technology, increasing work hours overall and dissolving the mental or social barriers workers put into place to separate a person's identity as, for example, mother or deacon from their identity as employee. Throughout the 2014–2015 school year, Du Bois's teachers supplemented the hidden curriculum with various ad hoc efforts to encourage presence bleed.

Du Bois didn't require teachers to hand out cell phone numbers, but efforts to build a high-performance academic culture still led teachers and students to use their PC's and phones to push the boundaries of the school far beyond its walls. This constant connectivity with her workplace was one reason Melissa quit teaching in February 2015: "There was no privacy." After her time in Houston, Melissa had tried to build more barriers between home and work, but the urgency of Du Bois's mission kept knocking those barriers down.

Other teachers embraced Du Bois's presence bleed. Constant connectivity was a sign of professionalism, and so encouraging students to practice constant connectivity was understood as a route to the sort of educational and career success that would secure a middle-class life for their students. Presence bleed, then, was not just a set of behaviors that encouraged a high-performance academic culture—checking email at night, reviewing one's grades in SchoolForce to inform the week's schedule—but also evidence of that culture.

Sam, physics teacher and senior team leader, embraced Du Bois over his old school in Philadelphia where he had taught for Teach for America—the teaching internship program wherein elite college graduates are placed for two years at underprivileged schools—because his previous institution "had little to no culture, no positive culture that was built by the school. It had its own culture, but it was because the kids ran the place and it was a pretty violent place." He told me Du Bois was engaged in a different, deliberate culture-building process. This meant believing in "the social justice mission of education" and taking pride in his college-bound students because "that's going to change life outcomes."

Encouraging this level of commitment in students required a great deal of commitment from teachers. Sam worried about the strain presence bleed placed on teachers, especially older ones, but embraced it himself because he figured, as a young man with no spouse or children of his own, this was the time to commit his whole self to a cause. "I would love to have fifteen emails a night from kids on homework," he said. And he answered them too, even if he had to correct them on salutations, subject lines, and the like. This commitment was similar to the way employees of InCrowd embraced the death of the nine-to-five workday as a way to fulfill the social mission of the firm and self-actualize in the process. Sam tried to call every one of his students' parents every three weeks, usually on his walk home from school.

This commitment showed up in other life choices, like Sam's decision to move to an apartment within walking distance of the school. Clara and Ryan did the same, fully aware that they were representatives of the same postrecession gentrification wave Clara's students studied in sociology. Ryan moved into Du Bois's neighborhood not just to minimize his commute or make himself more available to extracurriculars—like the school soccer team he coached—but because he and his wife—also a teacher—wanted to commit to being members of the community they were teaching in.

Some students embraced Du Bois's presence bleed. It was an example of how Martin said his teachers would "make students push to their limits." He appreciated teachers' willingness to connect outside of school because he went to work right after classes let out and was always doing homework at odd hours of the night or over weekends. Ryan said he saw evidence of working students' dedication in their constant connectivity—because the online platform he used to teach calculus kept a record of students' time in it.

Teachers estimated that around half of students lacked home internet access, though this number varied as people moved or bills went unpaid. For students without the motivation or support, this meant they simply could not do large portions of their homework at home. But engaged students without home internet access welcomed presence bleed, seeing it as a way to stay on top of their assignments whenever spare time and a Wi-Fi signal became available. Irene's mother worked nights cleaning offices, so when other students were finishing homework on their laptops, she was getting dinner ready for her younger sisters. That teachers would answer her emails on weekends was a huge help, because that's when she would bring an old Starbucks cup into the coffee shop, camp out, and use the Wi-Fi to access all the assignments, messages, and grades that were stored in the cloud. Eventually, she became such a regular that the baristas didn't even mind when she forgot the old cup she had been bringing in to pretend she had bought something. Of course, it wasn't all homework. Starbucks, especially once she got onto good enough terms with the baristas, was a rest space, like the library was for Mia, Ebony, and Shawn. With all her responsibilities at home and at school, sometimes Irene just wanted to take her school laptop to the coffee shop and get online to watch soap operas or murder mysteries. She was determined to make constant connectivity work on her terms.

At school, constant connectivity raised students' expectations of when and how teachers would communicate with them. Clara often used her time monitoring study hall[14] in the Think Tank at the end of the day to get on her laptop and catch up on grading and lesson planning. During the day, teachers' free periods were usually taken up by subject-team or grade-team meetings, so they were always desperate for a spare moment in which to work. In study hall, Clara was often approached by students asking when the grade for an assignment submitted earlier in the day—or even just a class activity or behavior grade—would be up on SchoolForce. Clara would sigh, rub her temples, and then, as her RECE training suggested, patiently explain her daily schedule so that students could empathize with her not having had time to grade since this morning.

On days I shadowed Amanda's Digital Music or Video Game Design classes, she would vent to me about the consumer mentality that she felt constant connectivity encouraged in her students. They demanded quick updates from her at all times and were disappointed when they did not get them. Principal Carroll labeled these students as low-functioning. They reacted to grade postings as though they were an alarm, asking teachers why they scored a 2 instead of a 3 and what they could do to bridge the gap. Students who were high-functioning would use SchoolForce as a prompt to adjust their goals and their strategies to reach them, building timelines for the rest of a project or a quarter.

High- versus low-functioning, in her usage, was less a marker of intelligence than of one's comfort with the rhythms of academe; unsurprisingly, this was often a marker of privilege, of the habits and knowledge imparted (or not) by middle-class parents (Calarco 2018). The more comfortable students were with SchoolForce, the more they reminded me of how InCrowd used Salesforce, the customer relationship management technology on which SchoolForce was based. Good workers and good students used the data they saw about their own performance and the aggregate performance of their peers to figure out how to match their pace to that of the organization.

But not everyone was keyed into this rhythm. Sam believed that every technology in the school—SchoolForce, laptops, phones, and more—empowered the top-performing third of students to connect more deeply and extensively with their teachers and their material than they could have otherwise. For everyone else, all the data points and communication channels available either distracted them from the task at hand or intimidated them to the point where they shut down.

Because technology worked differently for different students depending on their skills, engagement, and level of privilege, the practice of presence bleed could not fully succeed without the lessons imparted by the hidden curriculum. Constant connectivity mattered little if the connections made didn't support productivity. To produce a high-performance academic culture, teachers had to model professionalism *and* encourage constant connectivity. This was a lot of work.

Amanda's classroom was one of the few places where it all came together. Students in Video Game Design worked in teams throughout the fall semester to build pixelated side-scrollers about a social issue of their choosing: teen pregnancy, homelessness, college admissions, and so on. These issues often hit close to home, even if the structure of the final project did not. That summative assessment was a pitch to guest judges deciding whether to "invest" in their video game. I helped out as a judge. Another judge introduced himself by saying he had done the same thing students were doing on Monday in San Francisco, giving a two-and-a-half-hour pitch for a $1.5 million project. An audience member replied that they had just interviewed at Chipotle. Everyone laughed.

If a group member didn't show up to the final presentation, Amanda would remind the rest of them that "these are real life experiences, and if you missed a presentation in front of a major company" your career would be over. A professional dress code was rigorously enforced, and high-scoring groups were invited into a section of the room Amanda labeled the Executive Lounge, where they could enjoy the snacks judges were sampling.

There were surely elements of the hidden curriculum here, raced and classed visions of what technology professionals looked like; Amanda awarded extra points to students who wore sports coats, for example. But students bought into it in large part because they adored Amanda, a middle-aged former punk they nicknamed OWL (Old White Lady). They came to her riotously decorated room after school for advice, they bought into her cheerleading, they loved the subject. And this buy-in meant that they and their parents responded to her after-hours texts and emails, pushing themselves to get deliverables in on time and coordinate with their team-mates. But this extra work took a toll on her.

After the last Video Game Design class had finished their presentations and left the room, Amanda collapsed into a chair next to me. She put her hands over her mouth to stifle a tired scream. She was exhausted and talked about quitting at the end of the year to try something new, somewhere she would feel appreciated and get more positive feedback from superiors. She believed deeply in Du Bois's values, but she was tired of all those after-hours texts and emails, drained from all the work that had gone into planning and executing the video game pitch. What Melissa had experienced as an invasion of privacy—work reaching into home—Amanda experienced as overextension: the labor of keeping students constantly connected. Aman-da's students kept interrupting our conversation, popping in to hug her. She didn't blame them for her stress. She blamed the school's relentless focus on student grades and scores. The final presentations came right after the leadership team's ultimatum to senior teachers to boost seniors' scores. It weighed heavily on her.

Amanda's classes were far more focused on technology than any other teacher's. Her classes succeeded, in the terms of Du Bois's mission, because she deployed the hidden curriculum in a way most students bought into and because she encouraged the sort of presence bleed that kept students focused on the real curriculum. But it took a tremendous toll on her. That the pitch session made her want to quit teaching was a clear sign that Amanda's success couldn't easily be replicated. Not everyone was Amanda, and not every class was Video Game Design. And most students, like Sam said, did not have the privilege or motivation to manage constant connectivity in a productive way.

Throughout the fall semester, teachers had tried to build a high-performance academic culture through the implicit lessons of the hidden curriculum and the practice of presence bleed. But this was not enough. The numbers were still off, and new experiments were needed. Things changed for the spring semester, after the leadership team's ultimatum. To build a high-performance academic culture for the first senior class and reach the goals the school had to reach, Du Bois increasingly turned to SchoolForce and a practice of data-driven discipline. As a system covering the entire school, SchoolForce belonged more to administrators than individual teachers. This new discipline seemed to contradict the school's values. But the mission was at stake. One of the biggest obstacles to that mission, the thing teachers tried to challenge through SchoolForce, was the smartphone almost every student had in their pocket.

Bad Phones and New Discipline

Systems administrator Manuel quit working at Du Bois in February 2015 and moved on to the same role at another charter school. Teachers had repeatedly complained to him that the school intranet—built on a network supplied by the municipal broadband initiatives celebrated at the OCTO event in chapter 1—was unusable during the three periods when students were at lunch. In those periods, dozens of students whipped out their phones to stream videos or music, and the school Wi-Fi network would suddenly have twice as many clients as there were computers in the building. Manuel had offered a simple solution over and over: throttle bandwidth on mobile and/ or unregistered devices. But he was shot down every time because that sort of top-down disciplinary measure did not fit Du Bois's values. His work kept running into that conflict, and it was one reason he quit.

If he had stayed on through the spring, Manuel would probably have found those sorts of solutions embraced by his superiors. In the second semester, after the leadership team's ultimatum, teachers increasingly sought to realize the school's mission by disciplining student phone use. This discipline was largely carried out through the behavior-tracking features within School-Force. These practices often conflicted with Du Bois's values, but SchoolForce, like the school's mission as a whole, was beyond the control of any one student or teacher. When push came to shove, the system was used to satisfy the leadership team's demands for measurable progress.

"We gave you [students] a laptop for free and you treat it like it doesn't matter, but if I take your phone because you're texting in class you act like I just ripped your heart out of your chest!" Sam told me. And it was true. The work machine was never valued as closely as the machine that *could* be used for work but was usually used for something else. There were good reasons for this. Students often spent a large portion of their own income or their parents' income on their phone and their phone plan, whereas the laptop incurred a one-time, fifty-dollar deposit almost always paid for by parents. Du Bois also lacked a deliberate instructional space for computer use—partly because of an assumption that teachers would holistically integrate that subject into their curriculum, partly because something like digital literacy was never a subject covered by standardized tests, either the evaluative ones mandated by the state or the SATs, ACTs, and APs so crucial to students' college admissions prospects. As with the question of whether sleeping was permitted at MLK, Du Bois's mission clarified teachers' positions on phone use: snapchatting in class was unprofessional and thus forbidden.

What porn was for the library, phones were for the school: a small thing that threw doubt on the big project. Eventually, school infrastructure forced the issue. The SchoolForce student data system on which the leadership team depended for its bird's-eye view of student progress was used to create a new set of disciplinary measures that nudged students away from the phone and toward the laptop. New disciplinary measures resolved the ambiguous place of students' phones in teachers' work.

Every student had a smartphone, even students like Irene who spent some nights in a shelter. Phones surfaced class divides. iPhones were common but more expensive than Androids. Pay-as-you-go plans like the one Martin bought were a sign of living more paycheck to paycheck and/or without parental support. Irene's mother bought her a prepaid Android TracFone so she could check her school email at home and report in on her and her sisters' doings. In this way, phones stretched students' home lives and social lives into their school lives. Of course, most use cases were more playful than Irene's—these were teenagers, after all—especially so for students like Martin who figured out how to use proxies or other work-arounds for the Du Bois firewall.

For teachers, phones were the locus of in-class disciplinary struggles. They were distractions in the moment, and teachers' reminders to put them away could end up taking up large chunks of classroom time. Phones also symbolized a focus on technology not for academic gain or long-term professionalization but for momentary pleasure. The difficulty was that the school's values precluded a wide-scale confiscation program or other crackdowns on phone use.

Calculus teacher Ryan had happily left a traditional public school for Du Bois. The only time he missed his old DCPS job was when he could not persuade students to put phones away and he had no option to escalate. There was no option for confiscation or detention at Du Bois because that sort of discipline ran counter to the school's racial justice values. Ryan accepted it, working hard to build relationships with individual students so that they would respond to him and designating appropriate times for phones in class—googling definitions, looking up video demonstrations—so that students would understand that phone use was inappropriate the rest of the time.

But it was an uphill slog. He appreciated that many students "used [the phone] effectively when they've needed to" but knew it was "stressful [to be] a kid with a phone in school, trying to maintain a relationship or a presence online." And it was difficult for him to tell when the phone in the lap was being used to text message or check grades, whether they were checking out or, in Sam's words, "using their technology as a way to better themselves." Ryan kept the model of office work in his head as a way to work through this ambiguity: "I feel like at this point it's accepted, you have it out, and I'm going to try to teach you, like when you're a professional, to not have it out. And in most cases, the professionals that I work with don't have it out."

And so, the reminders to put phones away continued, but this forced a conflict with Du Bois's values. For two months in the fall, I tracked how often Melissa asked students to put phones away in her fifty-minute AP English Literature class at the end of the day. She averaged about ten reminders per class. This was a challenge within the RECE curriculum because each negative reprimand to a student was supposed to be balanced by three positive encouragements. Student phone use meant Melissa was constantly fighting to balance her disciplinary ledger.

The phone was where these racial justice values came into conflict with the charter's mission of human capital growth. High-achieving students, often from more privileged households, understood the technological distinctions teachers were trying to make. "I think that's why they gave us the laptops, to use it for good stuff and work and not use our phones—because we have social media on there," Daniella told me. But the majority of students, in Sam's reckoning, did not see this distinction and did not ditch their phones for their laptops during the school day. There were many reasons for this, ranging from stress to romance to disengagement.

Teachers tried to promote the laptop and the high-performance academic culture it represented through the hidden curriculum and presence bleed. But this failed to reach everybody. Students pushed back against the professionalization of their routines, seeking to reclaim spaces like the Think Tank as a safe space that provided the care they needed—whether that meant rest, socializing, or something else. Other students were just unengaged in this informal technological pedagogy, whether because of lack of interest, because they couldn't get online at home, or just because there was no room in the formal curriculum for digital literacy instruction. Teachers themselves had a hard time encouraging a high-performance academic culture and discouraging phone use: constant connectivity sapped their time and energy, and Du Bois's values prevented most disciplinary measures. Sam understood this as their failure, not students'. He feared the consequences they'd suffer "in a professional setting" for checking their phone under the table because "if you're a non-White person, the world is just dying to write you off."

In January, the leadership team made it clear that change was needed and that teachers needed to focus on changing students' habits to support a high-performance academic culture and get more seniors to graduate. So the faculty turned to SchoolForce, where a new set of Work Hard Grades was created. Work Hard Grades bridged subjects and thus made a student's habits visible to all their teachers, their parents, and interested administrators. Students were asked to designate a particular area they wanted to improve and be assessed on in their Work Hard Grades. With faculty's encouragement, phone use was easily the favorite choice.

In the second semester, individual teachers made SchoolForce data, Work Hard Grades and beyond, an increased focus of class time, training students to attend and react to their scores—disciplining themselves. At the beginning of the third quarter, Principal Carroll framed the changes as an opportunity for peers in homeroom to track each other's data and encourage each other to reach new targets. Before she quit, Melissa objected in a senior teachers' meeting—"That's a lot of SchoolForce tracking!"—but most of her colleagues were on board. Teachers approved of the plan in large part because it meant fewer disciplinary actions on their part. School-Force data would demonstrate the need for phone discipline, and students would carry it out on their own—rather than dealing, faculty hoped, with constant reminders from teachers. Phones had always been an obstacle to their mission, but their values prevented a firm stand on it. The leadership team stressed the importance of mission over values, and SchoolForce supported that move; not just because it put a hard number—the Work Hard Grades—on the ambiguous question of phones' place in the classroom, but because SchoolForce provided a mode of discipline that was both more comprehensive and more impersonal than teachers' individual efforts.

SchoolForce was an important piece of Du Bois's infrastructure. Pushed by the leadership team that won the grant for it and now kept a close eye on its use, it became the means to build a high-performance academic culture, overriding individual teachers' pedagogies and, to an extent, Du Bois's holistic approach to students' lives. Teachers never thought of their students as just a grade or a test score, but mid-year bootstrapping meant that those data points became increasingly more important parts of their pedagogy as the second semester started and seniors rocketed toward graduation.

Clara's sociology class was always the clearest example of Du Bois's values in action, engaging students in research on gentrification in their neighborhoods and, in the second semester, frank conversations about tracking, class reproduction, and systemic racism in education—including case studies of Du Bois! But by mid-March, just before spring break, even Clara had capitulated to the SchoolForce system for data discipline. One day she stood in front of a matrix on the whiteboard, each cell holding the passing rate for a given class. She highlighted the sociology classes and then showed the numbers dipping over time. Students whispered and whistled, surprised.

"We can do better than 57 percent," Clara said, and walked them through SchoolForce's presentation of their academic and Work Hard Grades once more. She encouraged them to look at this data and their upcoming assignments individually and with friends. They needed to ask themselves, "What is the thing I can do that can lift my grade?" Irene, sitting with me in the Think Tank's beanbag chairs during her free period, watched and whispered through gritted teeth: "I hate this school." Not every student agreed with her. Martin embraced the data-driven pedagogy, reasoning that students would of course resist direct instructions to put phones away because it just made the phones more attractive. Instead, "they should make it a game," giving students small rewards to shape their behavior.

Martin's observation was prescient. At the leadership team's encouragement, the data-driven discipline only intensified as graduation approached. Increased emphasis on presence bleed supported this approach: more emails, texts, and phone calls to students and their parents, reminding them of important deadlines or data points threatening their graduation. Data-driven discipline made clear that some forms of constant connectivity were verboten. Unprofessional phone use was penalized and trained away with some success; students got the message that their social media selves were not welcome in school. The increased reliance on SchoolForce meant that the students' school selves were increasingly visible, salient, and accessible to them and their parents at home and to other teachers and administrators who were not physically present when students delivered a particular presentation or committed a particular behavioral infraction. The system helped accomplish what the the hidden curriculum and the practice of presence bleed could not: producing a high-performance academic culture that looked a lot like the workstyles of the White professionals these Black and Latinx students saw at the front of their classrooms.

Conclusion: A Turnaround School

Sam believed there was a better way to teach digital literacy. "Right now, we are too much 'either you use it or don't use it at all,' and that's not adult either." He loved the moments when he went to reprimand a student for looking at their phone under the table but found they were googling a physics problem. To make more of those moments, he and other teachers were taking a much more prescriptive approach with their ninth graders. His ninth graders were told from day one when it was OK to have a phone or a laptop out, laptops that now lived in carts, and were rewarded or penalized accordingly. The sort of free rein the seniors had enjoyed in the Think Tank would not happen again; Du Bois had learned from those experiments. "They learned what to do with the younger kids by experimenting on us," senior Corrinne told me.

The experience of Sam's ninth graders would be the norm going forward. But there had been no time to waste for seniors, and ad hoc efforts like the hidden curriculum, itself often more of a reflex than a deliberate attempt, had proved insufficient. So the leadership team encouraged a series of technological and behavioral experiments that would help them reach their goals for the first graduating class. SchoolForce infrastructure nudged students away from phones and distractions and toward laptops and a high-performance academic culture. These were big changes. This focus on bad habits seemed to run against the values that made Du Bois Du Bois: restorative justice, a resistance to deficit thinking, a belief in student empowerment based on an explicitly political view of the needs of marginalized youth.

But change was necessary because the stakes were high. Anything short of sustained year-over-year improvements in test scores, GPAs, and graduation rates was harshly scrutinized by the municipal government, funders, and the leadership team. Du Bois bootstrapped because there was, quite literally, no room for failure. Sam, for example, had approached the leadership team to ask about plans for students who would need a fifth year of high school and was told to stay focused on the fourth year. Ironically, Ryan noted in one senior team meeting that the fifth-year plan was one place where DCPS shined. For all his criticism of the local public schools, he knew that they knew that there was no single path to success for their students.

Teachers told me they had no way to let students fail, reflect on that failure, and learn to live with failure as a regular part of the professional world. There were at least three reasons why failure was not on the educational menu. First, teachers expected students would pick up on their modeling of professional behaviors. Second, the staff's dedication to Du Bois's values meant they had a political and personal stake in pushing students to the finish line. And finally, the time and resources to practice certain types of professional failure just were not there. Digital literacy was not a testable subject, after all, so a PC basics class like MLK's was never going to happen.

The bootstrapping process brooks no failure. Too much depends on its success, for both the organization and the people it serves. Constant reconfigurations of habits and technology put everything up for grabs at Du Bois. Ryan said of his seniors, "They've always been guinea pigs and they know that." Irene said she and her friends were "little lab rats." As the rare senior who had attended Du Bois's elementary, middle, and high schools, she recognized that the school worked hard to make its values a reality. She would rattle off the school's mission—to get every student, regardless of their background, into the college of their choice—and she praised the way Du Bois's staff "do a really great job of giving you this support system."

She and some of her peers recognized that charters like Du Bois were, like startups, in permanent beta. The difference is that schools, like libraries, have political and economic limits placed on their experimentation because they cannot choose their clients but must serve the polity as a whole. Where InCrowd could shift from serving consumers to serving businesses, a public school in DC cannot choose to start serving, say, adults in Delaware. Charter schools, native to the neoliberal political turn, are further away from this democratic ideal than public libraries; indeed, limited regulatory oversight and greater flexibility in choosing their constituents (and vice versa) is part of their appeal—it's how they innovate. But even public-private partnerships are public to an extent, and the school's charter authorization hinges on a specific promise made to regulators: human capital growth, measured in test scores.

Du Bois went beyond this, casting the promise in racial justice terms and including any student who won admission through the lottery. All this meant that there were limits to how much experimentation could really change how the school worked and what it could accomplish. To the latter point, it has long been a stubborn fact of US education that out-of-school factors, such as family income and child nutrition, account for the majority of variance in student test scores (Goldhader, Brewer, and Anderson 1999; Sirin 2005; Tienken et al. 2017).

Still, experimentation continued. Their mission demanded it. School values, like everything else, were subject to revision based on incoming data. And teachers and administrators were constantly being presented with new data that showed the need for further bootstrapping. Such calls for organizational restructuring emerged not just from student performance data, but from outside assessments of the school's progress in its mission.

In late March 2015, the largely White leadership team and Board of Trustees met to hear an outside consulting firm's report on exactly why Du Bois's scores on practice SATs and the DC-CAS (the standardized test preceding PARCC) had fallen year over year. The consultants' presentation began with the dramatic announcement that this was a "turnaround school" in desperate need of big changes, despite one of the highest per-pupil funding rates in the city and high internal measures of teacher effectiveness and student satisfaction. Staff were resigned to the news. They pushed back a bit but recognized that those test scores and an anticipated 54 percent graduation rate were unacceptable.

The leadership team largely agreed with the consultants' observation that Du Bois lacked priorities because everything was a priority and that a "culture of nice," an implicit denunciation of RECE, lead to continued excuses for both students and staff. A new five-year strategic plan was rolled out. Everything would be fitted to this plan, or else it wouldn't be funded—staff and technology included. The board had confidence it could turn things around. The school would keep bootstrapping. Everything, including those prized values, was up for negotiation if it meant—as Amanda, the OWL, had put it to me—that they "make kids successful in the systems that we have in the world right now."

5 Bootstrapping

On a Saturday morning in December 2014, I attended the Education Innovation Summit with a school full of teachers, administrators, and entrepreneurs. We were hosted in the gym and classrooms of E. L. Haynes Public Charter High School in northwest DC, a gleaming, renovated charter school with classrooms filled with student laptops. Upon signing in, every participant had to consent to being photographed and filmed for publicity materials the organizers, the CityBridge Foundation and the NewSchools Venture Fund[1] (both education reform philanthropies), would produce after the event and during it on social media. I had been clued into the event by a Du Bois teacher who had been previously awarded a CityBridge fellowship. Everyone knew each other from similar events; both teachers and tech entrepreneurs were used to giving their weekends over to work.

Haynes Head of School Jenny Niles, soon to take a position as Mayor Muriel Bowser's deputy mayor for education, welcomed the 150 or so attendees—about a third of whom, by a show of hands, did not work in a school. She praised the ongoing partnership between charters and traditional public schools in DC and their shared vision of "schools as a unit of change" that could bring the "power of personalized learning" to high-needs students through skilled teachers, well-designed schools, and innovative use of educational technology. The new class of CityBridge fellows, all DC teachers experimenting with digital tools in the classroom, then took the stage to address the power of personalized learning. They compared their students' fates in the broken US educational system to the frustration of getting served the wrong drink at Starbucks, with the first fellow lamenting that "education is the only system that doesn't personalize." Their presentation ended with a slide full of different Starbucks orders, representing different educational services delivered by the same school.

Keynote speaker Jim Shelton, former deputy secretary of education for the Obama administration and current chief impact officer at educational technology startup 2U, continued the theme. He said that slide full of different Starbucks orders was quite a long way off and that educational technology had to fill the gap in the meantime. But he was hopeful, telling us that educators already knew the lessons about disruptive innovation and human-centered design that Silicon Valley was just now evangelizing. "You are on what is called the bleeding edge. It's a noble place to be."

Shelton received a standing ovation. Teachers were still discussing it in my next two workshop sessions: on free tools and practical classroom tips—"Take out your cellphones! Last thing you want to hear in a classroom, right?"—for tracking student data. He had praised educators and positioned them as powerful actors in need of tools that fit their ambitions, with knowledge the tech sector could learn from.

The lunchtime panel titled "Inside the Design Sandbox: Two Approaches to Next-Generation Schools" was a different story. The CEO of Matchbook Learning, Sajan George, opened by touting his credentials. As a consultant, he had moved from corporate turnarounds to school turnarounds, overseeing the transformation of New Orleans public schools after Hurricane Katrina (when every school effectively became a charter), as well as DC schools during the Rhee administration. He then started a charter management organization so that he could create his own schools, beginning in Detroit. My tablemates grumbled, and then complained more audibly in the next session, about George's glib approach to teachers' unions ("It's not something you want to start with...") and structural poverty ("from crisis to opportunity..."). Especially grating for these Black women was his comparison between the charter movement's persistence and the resolve Martin Luther King Jr. displayed in his Birmingham jail cell.[2]

The next speaker, Aylon Samouha, education design provocateur with Achievement First, was more warmly received. Still, few teachers understood what they were getting out of his pitches for his charter network's designs, given his honesty that his schools had money that most did not. But the audience nodded along with both speakers' core message: the public school model, born of the Industrial Revolution, was sorely in need of an upgrade. This required a fresh injection of digital tools that could personalize student learning and a fresh mindset born of experimental public-private partnerships.

There was no Education Innovation Summit the next year. Instead, City-Bridge invited entrepreneurs and teachers to Startup Weekend EDU: Next Gen Schools, where they collaborated on pitches for new types and brands of schools that could solicit investment, scale quickly, and deliver personalized learning.

Converging on Hope

The Education Innovation Summit was a moment where the cultural and organizational links between my field sites became visible. The same Gates Foundation that funded technology procurements for DCPL also funded CityBridge. The same agile management and venture capital investment practices built for startups were pitched to schools. Powerful men (in an audience made up largely of Black women) like Shelton, George, and Samouha moved with ease between these spaces, using the same message to unite them in the same mission. Everyone was clear that the problem of poverty was a problem of technology and, importantly, they did not necessarily need to believe wholeheartedly in that framing to endorse it. It was simply the best option on the table, the framework that best fit the high stakes of their jobs.

What had happened at the summit? Helping professionals, mostly Black women, working in public institutions, were united in their frustration with their organizations—out of date and over capacity. They felt, keenly, the needs of the people they had to serve, but didn't have the tools to serve them. In came outside entrepreneurs with stamps of approval from liberal foundations, disruptive startups, and the Obama administration. They framed the many different problems facing students as a single crisis of human capital deficits, said schools were unequipped to solve it, and promised new skills and new technologies as solutions.

Occasionally, teachers chafed at the dictates of this new gospel, knowing it did not necessarily fit the material reality of their work. But it was precisely because of that dire reality and their own collective commitment to public service that they were open to the language of the access doctrine, open to a project of fighting poverty through technology, of upgrading their schools in order to upgrade their students. They would commit to this new gospel because they had to; it made things make sense, helping the teachers to understand their overwhelming problems and find actionable solutions to them. They would in turn leave the summit and spread that gospel alongside the practical tips they picked up in workshops; that was the whole job of the CityBridge fellows.[3]

This chapter explains why this scene and scenes like it keep happening in US cities, drawing comparisons between the previous three chapters to explain what makes the access doctrine attractive and why places like schools and startups keep bootstrapping in an attempt to fulfill that doctrine, even as they acknowledge its repeated failures. Chapter 1 located the birth of the access doctrine in the neoliberal political revolution and its replacement of labor market interventions (i.e., the direct provision of jobs or money to those outside the labor market) with skills training, the climax of which was the Clinton administration's reframing of the problem of persistent poverty in the information economy as a digital divide. Even as the phrase itself became passé in scholarly circles, the spirit of the digital divide lived on in talk of digital equity, digital inclusion, and the like. Each revision complicated the original, binary, determinist story, but in so doing affirmed its power as the common sense of the day that must be reframed or rebutted. Were the idea of a digital divide not so powerful, there would be no need to critically examine and reexamine it, again and again.

Adopting the access doctrine as an operating philosophy requires a constant process of organizational change, modeled on the ideal type of the startup and its ability to pivot to new ways of doing business: bootstrapping. The remainder of this chapter explores the mechanics behind this process, using the descriptive work of ethnography to create an explanatory model of organizational change. This model has three nonexclusive pathways, explaining the push to bootstrap through (1) the generalization of helping professionals' narrow demographic experiences as a rubric for systemic change, (2) the bridging organizations that bring the tech sector into greater contact with schools and libraries, and (3) the stress that public institutions concerned with collective welfare face when their mission and the means to fulfill it are unclear, leading to an embrace of confident private sector models. This is what makes the problem of poverty a problem of technology for these organizations, which then go on to teach the rest of us to make those same links.

These pressures to bootstrap cross organizational fields, remaking not just individual professions or organizations but whole institutions of social reproduction. This is bigger than new language in new libraries, or new governance models in schools. The places that teach us how to make a living have been restructured around a new set of priorities. Politicians, entrepreneurs, and philanthropists have pushed schools and libraries to make new kinds of people on the model of the tech sector. These new kinds of people will, the access doctrine teaches, master the uncertainty inherent to the information economy, even if there are not enough tech jobs for them. This is a race-class project, in which the political coalition known in an earlier era as the Atari Democrats remakes cities in their image. Cities competing with each other for tax revenue welcome the change, as do schools and libraries desperate for resources, legitimacy, and clarity. Closing the digital divide becomes their mission because it ensures their survival.

Contained within each pathway to bootstrapping is a reason for its failure and thus the need to keep bootstrapping, each on display at the Education Innovation Summit. First, pathways to individual professional success often rely on sources of support outside those pathways (e.g., wealth, luck) and thus those individual stories cannot scale up, nor can they account for the larger, structural determinants of a training program's success or failure. Second, schools and libraries aren't startups and cannot suddenly change their mission or the people they serve; this is why I call what MLK and Du Bois do bootstrapping and call what InCrowd does pivoting. Third, making a problem more tractable does not solve the problem if its causes remain untroubled, and indeed strategic simplification can, by refusing certain needs of the marginalized, worsen the problem.

To explain the sources of the bootstrapping process and its repetition, I draw on neoinstitutionalist theory, which explains processes of organizational convergence—for example, why do so many new libraries look like Apple stores?—through organizations' search not for greater efficiency, but for greater legitimacy (DiMaggio and Powell 1983). A claim to legitimacy is a claim to one's rightful place in the social order. This is both a symbolic claim to a particular social role and a claim to the political, financial, or cultural resources needed to fulfill that role. There is a push and pull here between organizations trying to secure legitimacy and the people, offices, or institutions that can help them reach it—all with their own interests. Scott (1991) gives the example of a carpentry training program. It could be defined as occupational training, occupational therapy, or recreation, and each choice will imply a different set of relationships with state regulators, federal grant makers, unions, employers, trainees, professional carpenters, and so on. Each definition will pull the program in a different direction because goals, resources, and approval hinge on those outside institutional connections.

This process played out throughout my fieldwork as new modes of problematization required new tools and new relationships (Callon 1984). Defining the problem of poverty as a problem of technology reshapes the organization and its relationship to internal and external stakeholders. This chapter unpacks that process and what drives it: what incentivizes these changes, what makes them seem sensible and logical solutions to the problem of poverty. By connecting startup, school, and library, we see a phenomenon larger than the intra-field problems the the institutional isomorphism literature generally addresses. After defunding and delegitimizing institutions of social reproduction based on their purported unfitness for the information economy, the state and capital then coerce these institutions into solving problems that they didn't create and don't have the tools to address. Because the stakes are so high and the models of success in the tech sector are so unattainable, places like schools and libraries cannot help but fail in their duties. But because those duties are so important to their survival and that of the people they serve, they will inevitably keep trying. Even if the people served are further marginalized in the process.[4]

There are a variety of individual stakeholders in the bootstrapping process, with varying levels of influence on how institutional legitimacy is defined and measured. Some are internal to the organization: conflicted librarians, dedicated teachers, students and patrons making the space their own in the name of play, profit, community, or rest—at least until administrators take notice. Other stakeholders are external. These are the consultants who labeled Du Bois "a turnaround school." Or the donors and politicians invested in the library's future and its historic architecture, the "nice White people" for whom librarian Grant said he hated cleaning up the Digital Commons. These are the investors who could call for Travis and InCrowd to pivot. Each of these outside stakeholders exert pressure on the organization and help define its mission.

Organizations are not equally responsive to all stakeholders; indeed, one of the benefits of neoinstitutionalist theory is its specification of *who* matters in questions of organizational reform and *why* they matter. Many homeless patrons at MLK, for example, have a tremendous amount of political literacy, learned from years spent negotiating the complex bureaucracies governing their lives. But they rarely participated in hearings on the renovations. The glass walls of the Dream Lab were an important symbolic barrier. Cameron, a homeless activist who often worked out of MLK, said he noticed this pattern throughout the city government's consultations with the public. "I think that a lot of times city officials know the types of people that are going to be drawn to a certain meeting, and they just intentionally make sure not to reach across the aisle." The situation was similar in Du Bois school board meetings. All were formally open to the public, but outside attendees were rare. Students never attended, and there were never more than three parents at a given meeting, besides the parent representatives on the board. These parent attendees were, in the meetings I attended, always middle-aged Black women—mothers or grandmothers of students—with years of experience in the public school system or similar bureaucracies. Poor and/or Latinx parents were absent. Organizations that, like Du Bois, operate in permanent beta *can* make everyone's perspective valuable data, a constant input that changes operations. But they are rarely optimized to do so, because of routine, ignorance, austerity, or the simple fact that some stakeholders can reward them more than others. And so "there exists, unfortunately, a wide gulf between the experience of participating in the design of something and needlessly being subjected to instability—or being used for merely being a user" (Neff and Stark 2004).

Other institutional levers come in the form not of individual stakeholders or groups of them but administrative bodies that attempt to shape the entire institutional field of education or librarianship. For example, no parent carried the weight of the DCPCSB, which judged Du Bois's test scores and graduation rates and is ultimately responsible for approving any school's charter.

Other administrative bodies shape a field through funding. Both in life and in the foundation that he left behind, Andrew Carnegie, for example, greatly influenced the design, placement, and organization of libraries (Harvey et al. 2011) and art museums (DiMaggio 1991). Certain funding agencies play an outsized role in defining the problem of poverty as a problem of technology—the Gates Foundation or Walton Family Foundation, the Department of Education's Race to the Top grants program, the NewSchools Venture Fund. Libraries and schools desperate to both secure funding and to be seen as active, innovative organizations seek out these grants, resources captured either by the tech sector directly or the tech sector's way of thinking.

Finally, change within an organizational field can be prompted by professional and credentialing organizations, which shape the practices and ideologies of teachers or librarians. The power of the access doctrine, and the political coalition behind it, crosses these organizational fields, bringing professions like librarianship and teaching together under one new rubric. Recall Becca's complaint that the transformation of library schools into iSchools replaced service values with technical values or the message from the CityBridge fellows that the right sort of technological thinking could design the perfect, personalized education for every student—just like a Starbucks order.

Bootstrapping Pathways

I have sketched out some of the different stakeholders in the bootstrapping process, the different people and groups that have a say in how MLK or Du Bois pursued its mission of extending access to digital tools and the skills to use them, who prompted changes within the organizations when that mission was at risk. These organizations are more responsive to some stakeholders than others because of the resources and legitimacy they offer. It is now time to mark out these patterns of influence more explicitly. Why do schools and libraries try to solve poverty with digital tools and digital skills, remaking themselves in startups' image in the process?

There are three primary causes. First, there is the demographic and professional background of teachers and librarians. This is a *meritocratic model*. Second, there is the pressure exerted by professional organizations that bridge different organizational fields, thus bringing the tech sector into greater contact with schools or libraries. This is *technological professionalization*. Third, there is the uncertainty that public institutions feel when their goals and the path to them are unclear and successful examples from the private sector are made readily available. This is *mission ambiguity*. Table 5.1 summarizes these dynamics. Each pathway is essentially coercive—organizations do not receive the resources or legitimacy on offer if they do not take up tech sector models—but they are not often experienced that way. Rather, these simply seem like sensible reforms, best practices to help the people they want to help. The pathways are distinct but nonexclusive. For example, mission ambiguity can make technological professionalization more appealing.

Table 5.1
Causes of bootstrapping

Pathway	Source	Result	Limits
Meritocratic model	Demographics and career history of helping professionals.	Helping professionals believe the skills training that led to their labor market success is broadly replicable.	Training alone does not explain labor market gains, much less income or wealth gains. Metrics used to gauge organizational success are heavily influenced by factors outside the organization's control.
Technological professionalization	Bridging organizations present active professionals or professionals-in-training with tech sector models.	Technological skills, solutions, and organizational models are readily accessible and in some cases must be followed to secure funding and legitimacy.	Residual ideological and professional commitments to public service within helping professionals and their organizations.
Mission ambiguity	Schools and libraries are underfunded and overcapacity relative to what they are asked to do for students and patrons. Solutions to these problems are unclear, as are the organizations' ability to solve them.	Internally, focusing on skills training and technology provision clarifies overwhelming problems, making them tractable. Externally, these solutions demonstrate activity and legitimacy to powerful stakeholders who might relieve pressure on the organization.	Simpler problems are not necessarily easier to solve. Schools and libraries do not have the resources or flexibility to pivot like startups.

Meritocratic Model

The first and most obvious link between my three field sites is demographic: the middle-class helping professionals at MLK and Du Bois on the one hand and startup entrepreneurs on the other were largely White, middle-class, college-educated people who immigrated to DC in the hopes of a new job or for a specific, *hopeful* one. All had bachelor's degrees.[5] Many had advanced degrees, especially teachers (master of education) and librarians (master of library and information science), but also founders and nontechnical members of tech (e.g., Travis holds an MBA, Ji an MA in design).[6]

The skills-to-job pipeline appeared to have worked for them. While they understood that discrimination could interfere with the labor market, their experience of it seemed to reveal its natural logic. They worked hard, went to school, picked the right sort of training, and got good jobs. They felt they made the right choices and were rewarded for their hard work, and so it was natural for them to understand the labor market struggles of the people they served—whether in the present or future—as a problem of getting the right education, the right training, the right skills. They were primed to believe in the access doctrine's mission to help boost the human capital of those on the margins of the information economy. Why not? It appeared to have worked for the helpers. In a *Washington Post* profile, DC Public Library CEO Richard Reyes-Gavilan insisted that the mantra guiding him through the MLK renovation was that "libraries are not their buildings" but "engines of human capital" (Martell 2016).

These personal histories exert normative pressure within organizations so that librarians, teachers, and entrepreneurs identify the work habits adopted as they rose through college and their profession with the rise itself.[7] And the organization, implicitly or explicitly, models this for the people it serves; recall the way the hipster contingent became a proof of concept for the Dream Lab—just as much as any machine—and replaced older Black workers in the process, or how the hidden curriculum taught work habits to students at Du Bois. These were ever racialized organizations, wherein whiteness acted as an unofficial credential that both supported professionals' own advancement and convinced them that the people they served could do the same (Ray 2019).

The class composition of a profession—which is always also its racial and gender composition—thus supported the perception that what worked for one would work for most, even if inheritance, luck, investor interest, or other factors were more powerful than these work habits. In schools and libraries, these helping professionals' employment and educational histories differed from most, if not all,[8] of their students and patrons. This is a power relation inherent to the helping professions: the trained train the untrained in their image.

The meritocratic mindset has tight links to the class project that birthed the access doctrine. The Democratic Leadership Council from which Clinton and Gore emerged—alongside other centrist powerbrokers like Scoop Jackson, Richard Gephardt, and Joseph Lieberman—took control of the

Democratic Party in part by changing its electoral base. No longer would the party cater to a racially diverse working class and its unions; it would instead focus on mobilizing White suburban professionals. Most important among these were the so-called Atari Democrats, working in tech-sector office parks in areas like California's Silicon Valley or Massachusetts's Route 128 corridor (Geismer 2019).

Although not huge in number, these professionals were politically active, frequently profiled by the media (e.g., Wayne 1982), and formed the popular complement to the technology executives who helped shape Clinton and Gore's agenda. Distrustful of unions and high taxes, the Atari Democrats were often quite active in local civil rights struggles that assured equality of opportunity (e.g., fair housing), but resisted redistributive measures that affected them personally (e.g., local public housing and two-way busing; Geismer 2014). Working in nonunion fields like tech, in which professional certifications and advanced degrees were required for promotion, they built a culture of meritocracy into the new Democratic coalition. Importantly, the politics of individual teachers and librarians in Du Bois and MLK often diverged from this model. The conclusion will return to this point. But as reformers working to reenergize public institutions with legacies of union- ized employment, they bore the stamp of the DLC's now arguably successful race-class project.

It is worth noting that their educational experience was not even typi- cal for their own cohort and that some majority thinking—the tendency to identify your and your peers' experiences as representative of a larger social reality—was likely at play here. My informants in these professions were overwhelmingly (1) White and 2) in the first half of their postcollege careers, within the twenty-five to thirty-four and thirty-four to forty-four Census age cohorts. But only 36 percent of each of these cohorts and White Americans generally—it is remarkable how consistent that number is across these three different groups—have obtained a bachelor's degree (Ryan and Bauman 2016). Indeed, during my fieldwork, the BLS (2016) identified the most common educational experience of a twenty-nine-year-old not as having completed college and gone on to a steady career, but as having only completed some college and worked a series of jobs in fits and starts.

So even before they began to work with their students and patrons, the experiences of these professionals appeared to be unrepresentative of their own age and race cohorts. Their atypical biographies may have represented hope in mobility, even as it shaded their thinking about how exactly social mobility works. Irene seemed to catch on to this tic in her teachers' thinking when she said with her usual sarcasm that "apparently it's assumed that all White people are successful and whatnot." When she and her peers at Du Bois, or patrons at MLK, resisted top-down attempts to increase the space's productivity, they resisted the meritocratic mindset. They were building spaces focused on different values, like pleasure or safety or rest.

Teachers and librarians were often aware of this disjuncture between what worked for them and what might work for others. They knew institutional racism existed at a level beyond individual prejudice, even if they didn't know exactly how to talk about it or what to do about it. Physics instructor Sam, for example, read overwhelmingly negative comments on a *Washington Post* story about DC Public Schools students to his own class as a way of motivating them to hustle harder and prove the haters wrong: "The world at large has a negative perception of what you at your best is." Their cognizance of this gap in part explains why constant organizational change was necessary. The skills-to-job pipeline worked for them. They hoped to extend that opportunity to others and there was no time to waste. But on the other hand, they were, as well-educated liberals, aware of the structural nature of the problem and desperate to somehow solve it. It required an enormous amount of energy to try to close that gap; recall Amanda collapsing after students presented their video games. Nothing may ever be enough, but they had to try something.

Sam or Amanda's awareness of the structural constraints on student success hint at the limits of a meritocratic model. These limits are perhaps more visible in schools than libraries, given how intensely they measure the students they wish to succeed. Sam surely worked hard to get into Harvard and succeed there, but his parents' wealth and their PhDs surely helped prepare him for the experience. Given charters' special focus on students of color in urban school districts, it is sobering to note that Black households headed by a college graduate have 33 percent less wealth than White households headed by high school dropouts. Despite increasingly higher degrees of educational attainment and opinion surveys that show they heavily value education, "for black families and other families of color, studying and working hard is not associated with the same levels of wealth amassed among whites" (Hamilton et al. 2015, 3). Hamilton et al. argue that the assumption that educational attainment drives wealth accumulation is equivalent to assuming that carrying an umbrella causes rain: if there's any causal effect, it's more likely in the other direction.

Besides the long-term problem of wealth accumulation—assets owned minus debts owed, the sort of thing that allows you to launch a startup[9]—there is the more local question of the factors influencing educational achievement, typically measured by test scores. Decades of research show a medium to strong relationship between students' socioeconomic status and their academic achievement (Sirin 2005). Tienken et al. (2017) found they could accurately predict middle-school students' scores on standardized tests in math and language arts using only three community-level variables: the percentage of families in a community with annual income over $200,000, the percentage of people in a community in poverty, and the percentage of people in a community with bachelor's degrees.

This is not to say that teachers should just throw up their hands, quit their jobs, and become lobbyists or organizers to solve the real problems,—only that a narrow definition of education as a human capital enhancement mission will fail on its own terms because so much of its success or failure is determined beyond the walls of the training center. This can be difficult to see for White professionals, no matter their good intentions. Why wouldn't it be? Individualizing their success maintains the faith in both their own merit and that of their institutions.

Technological Professionalization

What stood out most to me about the Education Innovation Summit discussed in this chapter's opening was the general consensus of how to solve problems in education. Although many long-time public school teachers, primarily Black women, chafed at the business language on stage—sometimes explicitly antiworker—there was no real disagreement in any of the discussions I heard that schools were working on an outdated industrial model and that we could start fixing it through increased personalization achieved via the introduction of third-party educational technology (ed-tech) solutions.

Organizations like the CityBridge Foundation and the NewSchools Venture Fund promote this consensus, bridging different organizational contexts to spread the gospel of the access doctrine. These bridging organizations are not themselves libraries or schools. Rather, as grantors or professional organizations, they train cohorts of teachers and librarians, fund different schools and libraries, and, crucially, utilize the language and ideas of the tech sector. That language and those ideas may come from these bridging organizations simply sharing similar, high-level ideological commitments with the tech sector. They may also, more directly, come from foundations birthed from tech-sector profits (e.g., the Gates Foundation, the Chan Zuckerberg Initiative, Google for Education), from the direct employment of entrepreneurs who worked in or model their work on that of startups (e.g., the summit speakers), or by the simple cross-pollination that comes from training teachers or librarians alongside technologists and entrepreneurs. Men like Shelton, George, and Samouha may not have the specific domain expertise needed to improve a specific school, but their backgrounds in business and technology allow them to cross domains with ease and spread the message of reform (Sims 2017).

Whatever the source, these bridging organizations are engines of professionalization, presenting disruptive, tech-sector models to teachers and librarians, schools and libraries, that redefine what that work means. These models are attractive in part because their adoption is rewarded with different sorts of material (grants, training, technological infrastructure) and symbolic (awards, fellowships, degrees, public acclaim) resources. This then is the second institutional channel through which the pressure to bootstrap is exerted: technological professionalization.

DC's librarians, teachers, and entrepreneurs are often involved in professional organizations themselves or have their workplaces touched by bridging organizations that bring in ideas and tools from outside their field. These organizations help schools and libraries justify their mission to the public and enhance communication between members—thus exerting normative pressure on individuals' professional practice. Certain passionate members of the community, such as the CityBridge fellows, become institutional entrepreneurs who take on this bridging work as part of their passionate, personal investment in their career, seeking to make new kinds of schools or libraries (Tracey, Philips, and Jarvis 2011).

Many such professional organizations predate the neoliberal revolution and encourage an older public service logic. For librarians there is, after a year of employment, membership in the union, a local affiliate of the American Federation of State, County and Municipal Employees (AFSCME). Members of Eugene's hipster contingent were generally too new to the profession to learn its politics and participate heavily in the union (Eugene was a noted exception here, participating heavily) and so received more professional guidance from their iSchool networks or interest groups that also connected with civic-minded tech workers, such as Code for DC.

Other organizations are native to the neoliberal revolution, though it would be too simple to describe their emergence as mere epiphenomenon. iSchools are of particular interest here. Recall Becca's distress when the institution that granted her MLS degree changed its name from the College of Library and Information Services to the iSchool: "You're going to eliminate 'libraries' first of all and then you're going from Service to Science." There is a touch of projection here. *Science* is nowhere in the name. It is referred to as either the iSchool, like its peer institutions in the iSchool consortium or, more formally, the College of Information Studies. But the change was nonetheless real. These institutions have experienced rapid growth in the twenty-first century, often drawing new faculty from the user-facing side of computer science to grow programs in human-computer interaction (HCI) or from the infrastructural side of business schools to grow programs in information management.

iSchools emerged in the 1980s and 1990s as library schools closed, consolidated, or took a chance on expanding with a focus on information technology and information literacy. In their survey of the PhD degrees of current iSchool faculty, Wiggins and Sawyer (2012) found that only 10 percent of PhDs were awarded in library-focused fields and 11 percent in information-focused fields, while 30 percent were in computing fields. For librarians, this means their professional training inevitably touches not just traditional topics in the library sciences but ideas, resources, and people from computer science and electrical engineering, or management and policy (9 percent of faculty PhD degrees in Wiggins and Sawyer's study). There are new and different people in their classrooms and colleges, as compared to forty years ago. Technology training and business thinking thus became a greater part of the MLIS as part of the field's own efforts to evolve, survive, and grow.

This dynamic is more apparent with charter teachers in DC, who are, with the exception of a single school, entirely nonunion and often involved in interest groups around their particular subject area (e.g., local math teachers) or more organized professional associations such as Teach for America (TFA) alumni networks, as well as frequent professional development events like the Education Innovation Summit. We've discussed the effects of programs like the summit and the CityBridge fellows. A full history of TFA is beyond the scope of the present study, but it is sufficient to say that the organization was founded to recruit high-achieving, largely White students from top-tier universities to teach for two years in underperforming, low-income urban and rural public schools. Its promoters and funders are similar to, and often the same organizations as, the promoters and funders of other education reform efforts reviewed in the previous chapter. TFA's mission is to close the "achievement gap" between poor, Black and Latinx students and wealthier White students through local teacher placements that occur after a five-week summer training institute (Scott, Trujillo, and Rivera 2016). Both Michelle Rhee and her successor, Kaya Henderson, were TFA veterans. Prior to joining DCPS, Henderson worked in a series of administrative roles in TFA: recruiter, director of admissions, and executive director of DC operations (Osborne 2015).

TFA's recruitment process means that its summer institute, rather than a traditional education degree program, plays a large role in teaching its teachers how to teach. While Du Bois didn't employ TFA teachers, or any first-time teachers, plenty of its faculty had still gone through the program in years past and thus retained those values, as well as the professional networks through which TFA connects alums to opportunities for professional development, open jobs, and networking events. Ryan, the seniors' calculus instructor, and Sam, physics teacher and senior team leader, both came to the profession through TFA.

Ryan was a math major at Haverford before joining TFA to teach in DC. His first posting was the public high school down the street from Du Bois. Ryan was honest: "If there wasn't TFA, I wouldn't be teaching." He had learned about Du Bois's new high school because TFA had professional development events in Du Bois's middle school on nights and weekends. Sam pursued an interdisciplinary degree in social studies at Harvard before joining up with TFA to teach in Philadelphia public schools, saying, "I'm happy that I did it. It got me into teaching. I wouldn't have gotten in otherwise." But now its work concerned him, not just in classrooms but outside them:

> TFA basically tells you that you are amazing and that low-income kids need you and there's a lot of this thing that sets you up to feel like this sort of white knight that's coming in and doing great things. Then you start teaching and you suck at it and the distinctions in race and class between your students and possibly us, depending on your background, that's an issue.

I ended up working in TFA institutes for several summers. I worked as a Corps Member Coach for new members and a school manager for a whole school site. TFA has gotten a lot better at preparing corps members for the reality that they face, giving them some humility around the work that they do. I think that's good. I think the organization as a whole though is struggling with the idea of … is our job to give people a dose of classroom so they can then go and be a consultant, or something else? There is just an incredible pipeline of people who teach for two years, who taught in TFA and are now running stuff in education. There's a particular ideology, an ethos there. Some parts of that are troubling for me.

The ideology Sam is referring to is of a piece with the rest of the education reform movement: we need to take apart the broken public schools of the industrial economy and put them back together again in a way that will generate the human capital necessary for the information economy. But it is important to note, as Sam does, that this ideology does not function only, or even primarily, as a way to motivate young, White do-gooders to work for a few years at a lower salary than longer-tenured, better-credentialed local teachers. TFA trains a class of institutional entrepreneurs who move on from the classroom to connect their mission in local schools with work in educational policy or the private sector, making new kinds of social enterprises that bridge public schooling and corporate America. Or they go on to advance their mission in traditional electoral politics, educational policy, or through bridging organizations like CityBridge. In their interview study of TFA alums and corps members, Scott, Trujillo, and Rivera (2016) find that this broader social mission is a primary motivator for participants, over and above the local impact in particular schools or even particular school systems, and a better way of understanding the broad scope of the organization's impact.

The Broad Academy—one of the wings of the Eli and Edythe Broad Education Foundation, which also funds smaller philanthropies such as the NewSchools Venture Fund—performs a similar function for school administrators, training them in business approaches to managing people and resources. This class of institutional entrepreneurs then bridges the worlds of business and education. Broad shares with the Gates or Walton Family Foundation a general interest in human capital generation and the import of business language into schools: "Grants become investments, programs are ventures, and measures of impact generally involve the ability to scale up an initiative" (Scott 2009, 115).

As of 2020, the top three government officials in DC education—the DC state superintendent, the deputy mayor for education, and the schools chancellor—were all alumni of the Broad Center for the Management of School Systems. Kaya Henderson, deputy and later successor to Michelle Rhee as schools chancellor, touched several of what Scott (2009) calls *venture philanthropies*. In contrast to the more broad-based, infrastructural approach of earlier foundations such as Ford, venture philanthropies fund a variety of technical, administrative, and curricular experiments, with the expectation that most will fail while a few will generate a significant return on investment—generally defined in terms of test scores—and demonstrate the possibility of scaling up. Since leaving DC city government, Henderson has been a fellow with the Chan Zuckerberg Initiative (Facebook founder Mark Zuckerberg's foundation), a superintendent-in-residence with the Broad Academy, and now head of community impact for Teach for All, TFA's global education portfolio.

These professional networks create normative pressure for homogenous practices *within* organizational fields (e.g., among fellow librarians, teachers, or techies). And then certain of these networks create similar normative pressure *between* organizational fields, so that librarians and teachers begin acting like entrepreneurs; although it should be noted that because these models often come with rewards for following them, these normative pressures can have an element of coercion. iSchools, in their curricula and their alum networks, build bridges among tech, business, and the library. NewSchools does this work for schools. Certain philanthropic ventures from the Gates Foundation or Google (e.g., Grow with Google, Google for Education) play this role for both schools and libraries.

You can also see evidence of this interorganizational linkage in the professional backgrounds of charter school administrators and trustees. Du Bois' board of trustees included two ed-tech executives (one of whom incubated their product at Du Bois), a biotech executive, a local neighborhood commissioner, two executives of national nonprofits, two senior leaders at major education reform philanthropies, three legal professionals, an executive director of a major labor union, a finance professional, and an education consultant who has moved between city governments and organizations like TFA. Lessons about bootstrapping are not delivered solely through helping professionals' training but through the networks that training creates and through the governance networks that provide support and resources for the flurry of organizational change necessary to meet their social mobility mission. And whether it's through a short-term fellowship or a longer-term career that leads into business or policy, librarians and teachers passionate about their mission may themselves become institutional entrepreneurs who teach their peers to bootstrap.

In their ambiguous reflections on their own training, Sam and Becca each testified to the limits of this pathway. Both hold to an older public service ideal wherein schools and libraries are vehicles not for human capital enhancement, the training of future digital professionals by current ones, but for the sort of collective welfare that does not necessarily generate returns in the labor market or any other market. This ideal school or library does not have to have actually existed in reality in order to inspire teachers or librarians, just as the ideal startup does not have to have actually existed in order to inspire entrepreneurs or those jealous of their success. It is a horizon upon which mission-driven individuals set their sights and against which they judge themselves. This was also visible in librarians' ambiguous reflections on porn. Clearly, finding porn is not a marketable skill—even though dodging the various filters and staff required a good deal of digital literacy—but for librarian Rachel, "If they don't have access to a computer in another spot and it's an outlet, it doesn't really bother me." That many teachers and librarians often approach their job as a calling to public service limits the degree to which technological professionalization can reframe the nature and purpose of those jobs. Every story from every helping professional in this book is riven with this sort of ambiguity.

This public service ideal is not only a matter of a vocational calling. Limits to technological professionalization also emerge from the publics being served and from the rules governing the institutions of public education and public librarianship. Patrons frequently reclaimed the Digital Commons as a place for games, for unsanctioned entrepreneurship, and, most importantly, as a free and open place to find respite from the indignities of a gentrifying city. Rest, and sleep in particular, emerged as a frequent site of conflict over who could use the library and for what—a conflict that could be diffused into different spaces but never fully resolved. It was a site of conflict in part because other means of refusing service to nonviolent patrons were illegal. That DC libraries can only kick out sleeping or violent patrons is in large part due to the 2001 lawsuit brought by homeless patron Richard Armstrong against the system's 1979 rule to kick out those with "objectionable appearance." A federal court of appeals ruled that this violated Armstrong's First and Fifth Amendment rights.

Schools similarly have their public service ideals codified into law, even as long histories of segregation and recent neoliberal innovations show that the ideal is more important to organizational identity than organizational practice. Traditional public schools must welcome all neighborhood students of appropriate age through their doors. Du Bois followed this lead, though of course students had to take the extra step of winning entrance via lottery. Open admissions fly in the face of the human capital frame. Not every investment will bear returns, and so the portfolio must be tightly managed. This may in part explain why over the course of a normal academic year DC charter schools tend to lose students, while traditional public schools tend to gain students, concentrating low-income students in traditional public schools and stretching the resource capacity of those schools—the funding for which is decided on a per-student basis before classes begin (Roth and Perkins 2019). No school can pivot like a startup, but charters can bootstrap better than traditional public schools.

Mission Ambiguity

Finally, schools and libraries face a grave responsibility under the access doctrine—resolving a human capital crisis—but the way to meet that goal is not always clear. This mission ambiguity prompts a search for models, such as those offered through technological professionalization, which are adopted to resolve the ambiguity internal to the organization and demonstrate activity and legitimacy to stakeholders external to the organization. When faced with overwhelming problems—child poverty, homelessness, the achievement gap—and an unclear pathway to their resolution, the comparative success and optimism within the tech sector becomes attractive, particularly when those models are readily available and frequently rewarded.

When Shawn said he'd "always been a computer man," he was expressing a complex knot of needs, relationships, and aspirations, converging at the computer lab. The Digital Commons was for him, and many others, not just an educational space, but a social space, a play space, a rest space, and more. Overstretched librarians like Grant understood this complex reality, but found it difficult to respond in kind—they did not have the time, training, resources, or institutional remit to provide everything homeless patrons needed or wanted. Messages from the municipal government like

"The internet: your future depends on it" were comparatively simpler, and carried not just political legitimacy but the promise of additional resources. It's no surprise those bootstrapping messages won the day and came to dominate the institution's thinking.

The normative pressure exerted by the racial and class identities described earlier—where social mobility through skills training makes sense to a specific cohort of White professionals—here intersects with a clear mimetic pressure: schools and libraries understand their charge but not how to pursue it, and so they seek out models. Even the operationalization of that mission into specific goals is a matter of much debate (e.g., librarians' uncertainty over the internet porn question, teachers' contentious relationship with phones). Turning the problem of poverty into a problem of technology resolves this ambiguity, creating a series of actionable if-then proposals: learning to code leads to jobs, training to code leads to funding.

At school and in the library, the practical definition of success was contested. While Du Bois was dedicated to getting every student into the college of their choice, there was much debate, as the first class of seniors approached graduation, as to whether that was fair to each student and their hopes and dreams. That some students wanted to proceed directly to the workforce did not fit Du Bois's definition of success. Debate over what success meant in the library played out in its spatial politics, rearranging where people and machines went and how they were connected. When pressed, teachers and librarians always admitted that there are no easy answers to what they can and should do about poverty. They saw the scale of the problem every single day and were for the most part humble about their own power. But they needed answers in order to continue their work. Librarians must know who gets more or less time on a computer. Teachers must know what to do about the student texting in class.

Examples are thus extraordinarily valuable. They suggest scripts for day-to-day operations and pathways to prove legitimacy and activity to powerful outside stakeholders such as parents, donors, and state review boards. Such examples are forthcoming from the tech sector, whether those examples are found in the media, via bridging organizations, or within the workplace. These examples might occupy the same space in which librarians and teachers work (e.g., the Dream Lab coworking spaces or the DC Tech demo series that brought hundreds of entrepreneurs to MLK every month). They might come from training spaces shared by both sectors (e.g., iSchools). Or they might be presented to personnel by powerful members of their professional networks (e.g., the Education Innovation Summit keynotes). These hopeful examples of social mobility—for individuals and organizations—inspire organizational changes meant to mimic those examples, even if, as most tech insiders readily admit, the majority of startups fail and shut down in the first year—something most would consider unacceptable for schools or libraries.

These examples are especially important in times of institutional flux, uncertainty, and looming illegitimacy. Recall the budget crisis for DC Public Library in the early 2000s and the clear messaging from library leaders today, especially Reyes-Gavilan, that the flagship central branch was a national embarrassment, positioned awkwardly in a now-booming downtown commercial zone. Or the pressure teachers and students at Du Bois felt to get to graduation because that first class of seniors were prototypes or, in Irene's words, "lab rats" for the school's mission. At the end of the year, consultants delivered bad news about the results of that prototyping process.

With this sort of pressure bearing down and the organization's purpose and identity at stake, following outside examples often means following the money and the political support that will keep an imperiled organization alive. Local pressure is often based in the mundane politics of city budgets, test scores, and comparisons with more well-heeled peers. For schools especially, those politics are grim. The US federalist system, combined with long histories of residential segregation, means there is significant inequality between school districts and between different states with different funding formulae (e.g., the relative importance of local property taxes versus state funds; see Moser and Rubinstein 2002).

Venture philanthropists step into the financial gaps created by these inequalities and deliberately widen the legitimacy gap between the public service institutions of the industrial economy and the bootstrapping institutions of the future. In a 2005 op-ed in the *Los Angeles Times* entitled "What's Wrong with American High Schools," Microsoft founder and Gates Foundation cochair Bill Gates, now arguably the most powerful man in education policy, wrote:

> Our high schools are obsolete.
>
> By obsolete, I don't just mean that they are broken, flawed and underfunded—although I can't argue with any of those descriptions.

What I mean is that they were designed 50 years ago to meet the needs of another age. Today, even when they work exactly as designed, our high schools cannot teach our kids what they need to know.

Until we design high schools to meet the needs of the 21st century, we will keep limiting—even ruining—the lives of millions of Americans every year. Frankly, I am terrified for our workforce of tomorrow.

Gates went on to argue that it's not just tools or curricula that are broken, but schools' expectations. He supported extending the most challenging curricula to all students, something like Du Bois's AP for All program. He proposed a program consisting of national graduation standards for workforce readiness, the relentless measurement and publication of school performance, and the takeover or shutdown of "failing" schools. Venture philanthropists like Gates are happy to intervene in local politics to ensure their examples are followed. Michelle Rhee raised around $200 million from private donors to fund a broad restructuring of how DC hired, rated, and promoted teachers (Biddle 2009). When Adrian Fenty lost the mayoralty to Vincent Gray, foundations threatened to pull $64.5 million of that funding unless Rhee was maintained (Turque 2010). Kaya Henderson's promotion satisfied them.

Once it is judged on its productivity, education becomes an immensely risky endeavor. Maintaining funding for everything you want to do and making a plan for student success that will, against all structural barriers, not just bear fruit years down the road, but bear fruit that can be measured and scaled up—this is difficult, uncertain work. Uncertain work in an environment where the template of public education is not only underfunded, but repeatedly delegitimized. These precarious organizational conditions force schools into entrepreneurial fundraising cycles: schools are starved of political and economic resources, they see venture philanthropists naming their funding priorities, and so schools reshape themselves to meet these priorities in order to get funding (Lipman 2015). Indeed, Du Bois had dedicated fundraising staff for exactly this purpose, and despite the school's racial justice values, it was not hard to see top philanthropic priorities throughout the school—especially in its heavy use of educational technology and its focus on year-over-year improvement in test scores.

US public libraries have a long history of engaging with philanthropy, going back at least as far as industrial titan Andrew Carnegie's library-building mission. Today, the successor to Carnegie's mission is, again, Gates, whose foundation funds grants to provide libraries with computers and internet infrastructure in an attempt to close the digital divide. There are important parallels to Carnegie's mission to subdue labor unrest through the dispersal of "useful knowledge" and Gates's mission to wire every library in order to connect patrons to job opportunities and education (Harvey et al. 2011; Stevenson 2010). Postrecession slumps in state and local budgets meant that alternative funding sources, especially philanthropic ones, became increasingly essential for keeping public libraries afloat (Agosto 2008). Stevenson (2009) argues that in technology policy specifically, this results in public libraries adopting the Gates Foundation's definition of the digital divide and its solutions for it, over and above what librarians themselves might prefer (e.g., proprietary rather than open-source software becomes the default).

Even if they're not involved with fundraising, professionals find themselves repeatedly presented with the example of tech companies that garner the sort of support from state and capital that teachers and librarians can only wish for. The school turnaround expert who raised some teachers' hackles at the Education Innovation Summit had had a successful corporate career, then a successful career as consultant to public school districts, before finally trying his hand at charter management. At the same event, Jim Shelton framed his disruptive ideas in the context of his move from deputy secretary of education to 2U's chief impact officer—after a year in which 2U produced $83 million in revenue (though no profit) and had a very successful initial public offering that left its valuation at approximately $1.4 billion (Kolodny 2014). For relatively lower-paid, overworked professionals in libraries and schools in search of clear answers to overwhelming problems, these examples are not only available but meaningful.

The limits to this pathway are similar to those for technological professionalization. First, simpler problems are not necessarily easier to solve—particularly when the source of the problem sits outside the organization. For example, while a focus on professionalization may resolve the question of whether or not someone can sleep in the library, it does not change the fact that many, many people need a place to rest during the day and that there is little public space remaining in the city, particularly during extreme weather, that they can enjoy. Reducing ambiguity resolves personal or organizational questions about how to proceed—which is important, the work must continue—but that is not the same as solving the problem.

Second, while tech-sector examples are attractive, schools and libraries have different constraints on their mission. They can't pivot like startups. This contrast is clear even just in the three organizations studied here. Where the environment of economic uncertainty pressured each one of my field sites, only InCrowd could pivot with confidence. Its mission was comparatively simple: sell subscriptions, increase revenue to the point that going public or being acquired by a larger rival becomes viable. This is not easy, by any means. A lot of blood, sweat, and tears are involved. But despite the difficulty, there is a clarity to the end goal that makes the pivot easier. Increasing revenue and investment provides the resources to cushion the pivot. Unprofitable products can be dropped, employees can be laid off.

In contrast, schools and libraries often have non-negotiable success metrics imposed by external stakeholders (e.g., graduation rates, test scores). Although philanthropies can set organizational agendas under austerity, their investments do not generate financial returns that can be put back into the organization, nor will they ever match the sheer scale of the investment capital seeking returns in tech. And of course, while InCrowd was free to pursue fundamental changes to its business model from consumer to enterprise software, the same level of flexibility is not possible in schools and libraries because of public expectations and institutional constraints. Even a charter school will not, over the course of an academic year, shift its focus from STEM to language immersion; too much internal and external consultation is required. Despite the widespread acknowledgment that a renovation was needed, MLK's took years to approve—much less fund.

Conclusion: Competing Institutional Cultures

The neoliberal revolution reframed the human cost of deindustrialization and the information economy's increasing labor market bifurcation as an issue of human capital deficits, resolved on the one hand through the education, technology provision, and deregulation that extends opportunities for competition and on the other hand through the expansive carceral solutions that punish or cage those who refuse to compete correctly This chapter showed why this story is so eagerly embraced by the organizations fighting poverty, themselves increasingly imperiled by the long-term effects of a shrinking welfare state and the more local effects of urban austerity. While they readily admit that the problems they face are much bigger than new

tech or new training can address, that story and the concomitant restructuring it demands garners these organizations new resources, new political
legitimacy, and new clarity in their operations. To train the workforce of
tomorrow is to assert that there is no tomorrow without those trainers.

The access doctrine cannot succeed as propaganda, no matter how per-
suasive it might be. For this sort of political common sense to take hold,
it must become a part of everyday life, explaining the present historical
conjuncture and offering a path to the good life within it. Drawing connections among startups, schools, and libraries, we have found three distinct but overlapping pathways whereby schools and libraries and the
people working within them turn the problem of poverty into a problem
of technology and begin the process of rapid and repeated organizational
change—bootstrapping—required to act on this problem definition. First,
the racial identity and professional history of helping professionals leads
them to empathize with the skills-training mission because it appears to
have worked for them. Second, bridging organizations spread these ideas
and bring models from the tech sector into schools and libraries. Third,
the scale of their mission and the uncertainty inherent to both that project
and the general economic environment in which it is pursued make the
examples provided by the tech sector more salient.

Bootstrapping, then, does not just name the changes that these organizations enact based on the model of startups. It also names the theory of
change within these organizations, the motivations driving these changes
and what results participants in them hope they will bear. We have also seen
that each pathway to bootstrapping comes with a set of limits that con-
strain organizational possibilities, particularly in comparison with the ide-
alized flexibility of startups. Many of these limits emerge as a result of older
public service ideals held by teachers or librarians, enforced by students or
patrons or built into the legal and physical infrastructures of schools and
libraries. It is precisely because bootstrapping is a cultural force, and not
just a recipe to follow, that each time organizational aspirations butt up
against these limits, a new round of changes is pursued. The stakes are too
high, both for the organization and the people it serves, to do otherwise.

With this in mind, I want to close this chapter by suggesting that bootstrapping is a new institutional culture, a supraorganizational script for the
work of social reproduction: how to make people and teach them to make
a living. This institutional culture is practiced by teachers and librarians

at the level of individual organizations, providing both opportunities for action and limits to them. It is communicated and endorsed by people and groups that move between these organizational fields, providing legitimacy and resources to schools and libraries. And it provides a script for many different schools and libraries, a story about what *the school* or *the library* as an institution should be in the twenty-first century.[10] It is preached by the powerful representatives of larger social structures—witness Clinton's message that "what you earn depends on what you learn"—but this does not obviate the individual agency of these teachers and students or librarians and patrons (Ifill 1992). Rather, bootstrapping provides a way to understand how their individual choices are empowered or hindered by particular organizations and their particular political-economic location. Here, the work of making people and teaching them to make a living shifts in response to broader changes in capitalist production

It is important to define bootstrapping as a new institutional culture in part because limits on bootstrapping emerge from a different, older institutional culture centered on public service and collective welfare. We saw it in action whenever librarians justified the sleep or porn they knew was not permitted in the computer lab, or when teachers resisted the discipline they were increasingly encouraged to visit on students and their phones. This institutional culture set a mission of public resource provision beyond the vagaries of the market, with the organizations themselves built to last in the same form for many decades because that mission would never go away. Public service organizations are shock absorbers for predictable economic downturns and stewards of a national cultural vision—however exclusively or violently that is defined in practice—that extends beyond any given cycle of recession and recovery. They are archives and safehouses, points of public care built as public goods. Their bootstrapping successors are training centers and resource banks, targeted investments built of public-private partnerships. The two competing institutional cultures, as observed at MLK and Du Bois, are summarized in table 5.2.

This typology likely applies to other institutions of social reproduction that have been similarly remade through information technology in the service of the access doctrine: social work, higher education, community development, and so on. Both institutional cultures are ideal types, visions of what these organizations should be. No single organization is completely run on one rubric (Besharov and Smith 2014). Bootstrapping and public

Table 5.2
Competing institutional cultures

	Public service	Bootstrapping
Political era	Keynesianism	Neoliberalism
Funding streams	State and local taxes	Diminished taxes, supplemented by competitive grants
Orientation to peer organizations	Nodes in state and local networks	Inter- and intraregional competitors
Orientation to people served by the organization	Safe space	Training space
Orientation to technology within the organization	Public good	Vector for competition
Orientation to poverty	Provide support	Provide skills
Sources of practitioner training	Colleges of education and library science	Colleges of education and education-preparation providers such as TFA; iSchools
Pace of organizational change	Regular, usually by state mandate	Frequent, as a matter of survival

service necessarily overlap in practice. People take what they need from each script (Friedland and Alford 1991). Because the former succeeds the latter, bootstrapping necessarily builds from the materials and practices of public service organizations; recall that Du Bois was literally built inside a shuttered public school.

And in practice, each vision is compromised by political struggle. Long histories of racial segregation make it clear that rarely has the public service institutional culture in schools and libraries actually served *everyone*. Today, schools and libraries and similar spaces continue, in many cases, to be sites of discipline and violence against some students, especially working-class Black and Latinx students (Shedd 2015). Public service might motivate some of this, as an attempt to "save" these kids—one source of Sam's uneasiness with TFA as discussed earlier—and it might motivate other teachers to create safe spaces for them, as Clara and Principal Carroll so often did.

Nor are these institutional cultures only compromised from above. The daily resistance of students and patrons was one thing that prevented the bootstrapping mission from being realized in Du Bois and MLK. Their resistance was not necessarily based in a philosophy of public service; rather, they took what they wanted and needed so that they could flourish within these bootstrapping spaces. It is important to recognize the political power of students or patrons and that this power does not necessarily comport to the terms of White professionals' own internal struggles. Rather, students and patrons strategically utilized one institutional culture or another at different times to elicit support from a teacher or librarian or to avoid their scrutiny.

Nevertheless, bootstrapping provides a powerful driver of organizational change, replacing an older public service culture in the process. It is here that the problem of poverty becomes a problem of technology for the people who are trying to solve that problem. It is here that the access doctrine is taught and learned and circulated within these organizations. It is here that the terms of social reproduction are reset, with schools and libraries following startups' examples and passing them on to the people they serve. It is here that we learn that we must learn to code, or else. And it is here that the people whom these institutions were built to serve are further marginalized in the process, because these institutions must bootstrap to save themselves.

In this way, bootstrapping resembles processes of *predatory inclusion* observed in settings like consumer credit and for-profit education: the extension of long-withheld opportunities or resources to marginalized groups who seek social mobility, but on terms that disadvantage them in the long term and eventually reproduce inter-group inequality (Cottom 2017; Seamster and Charron-Chénier 2017). But unlike subprime home loans or for-profit college degrees, the opportunities offered on the terms of the access doctrine do not only marginalize those seeking social mobility, they harm the very institutions offering those opportunities. Organizations like schools and libraries may find momentary security through bootstrapping, but only if they are prepared to sacrifice their values, overextend their employees, accumulate debt, change their mission, and revise any other piece of their identity or operations as necessary. The stakes of their mission are so high that every option must be on the table. In this way, it is not just that welfare state institutions within neoliberalism prey on the people they serve, though that does happen, especially in contexts where revenue can be generated or where administrative capacity has been shifted to more carceral ends (Gilmore 2007; Gilmore and Gilmore 2008). Rather, in the process of including the marginalized and offering them access, organizations like schools and libraries will often prey on themselves; an exercise in self-cannibalization meant to appease the external stakeholders on which desperate institutions rely and to motivate the internal stakeholders from whom so much is demanded.

The access doctrine emerged from a particular political project at a particular moment in history. Deindustrialization, the defeat of organized labor, and the violent substitution of the welfare state with the carceral state—all are engines of immiseration, the human toll of which stubbornly persisted in the supposed plenty of the information economy. All were simplified as a digital divide between those with the human capital necessary to succeed and those who needed the right tools and skills to reach that level. Next, we return to that structural perspective, above the everyday fray of organizational life, to conclude this investigation and reflect on the political purpose the access doctrine continues to serve within this particular stage of capitalism.

Conclusion: Reproducing Hope

In March 2017, I was living and working in Boston but knew I had to return to DC to witness the long-delayed closure of the Martin Luther King Jr. Memorial Library in support of its long-anticipated renovation. Librarians and patrons had kept me updated on the starts and stops in the renovation timeline. But a date had finally been set. I arrived on March 3, the day before the closure. Certain collections, offices, and technologies were being emptied out, but daily life continued apace. Everyone knew the library would reappear, but the uncertainty over when that would happen and what its new form would mean for this fixture of downtown DC meant there was an air of unease in the building, now interspersed with photographs of its 1972 grand opening. It all had the feeling of a funeral reception.

Patrons and librarians were going through the motions, but the building felt empty and no one was quite sure what to say to each other. I walked in just as four police officers were pulling a Black man, hands and ankles cuffed, surgical mask on his face, out of the front doors and into a squad car. He was wearing a t-shirt in the early spring cold. A patron explained to me that there had been a fight, but the front desk would only tell me he had been "disrupting the library."

The Digital Commons filled up over the course of the afternoon. The Dream Lab's glass cubicles remained empty. Handmade signs hung over submission boxes asked patrons to leave stories about the library. A couple of members of Mia's crew were still hanging out, playing video games and cuddling. At the end of the day, a librarian announced, "The library is closing in thirty minutes. Your computers will shut off in twenty minutes. The library will reopen at 9:30 tomorrow and then close at 5:30 for a long time." A sleeper next to me perked up and asked: "It's closing?" His neighbor and I walked him through it. Shortly, the police entered and repeated the same announcement a little louder. Librarians had flyers ready for homeless patrons, mapping out the various churches and shelters downtown in which people could find refuge while MLK was closed, as well as the branch libraries that were a good walk or a short bus ride away.

The flyers noted that DC's Department of Human Services (DHS) was expanding services at the Adams Place Day Center, opened in 2015, and adding drop-in hours and computers to replicate the MLK experience. A new shuttle would take shelter residents there, or to the intersection of Minnesota Avenue and Benning Road NE, where an employment center, a health clinic, and a branch library were clustered together. Neither option was as centrally located as MLK. Adams Place was on the northeast edge of DC, on a strip of road where the city government pushed loud or hazardous businesses. The Minnesota and Benning NE intersection was in a majority-Black neighborhood east of the river that was only beginning to feel the disruption of gentrification. There was nothing like the restaurants, stores, museums, and businesses surrounding MLK. To their credit, librarians, at the urging of patrons and community groups, had mobilized to make patrons aware of nearby resources. But librarian Eugene said they were frustrated with the lack of planning on the city's side, where DHS seemed to realize the crisis of effectively closing a hundred-person day shelter a little too late. It remained to be seen whether other branches, especially nearby Shaw, were prepared for the new patrons they would receive.

I returned on closing day, walking through front doors now covered with renderings of the future library. Librarians offered tea and snacks. Patrons took selfies in front of the mural of Dr. King. Parents still went to story time. Rules of behavior were still posted next to every elevator. Police still rapped their knuckles on the table next to sleepers. Upstairs, across from adult services, there was a big handmade map of DC that asked patrons to write what they wanted out of the city's future. "Less gentrification" was biggest. Nothing was written on the neighborhoods east of the river besides "better schools." Disabled patrons were getting every last minute of use out of the Adaptive Services Lab. Down in the Digital Commons, I said hello to Eugene, who would now start working in the DC Jail. A police officer walked up to the librarians' desk. The librarian said, "It's about to close." The cop replied, "Can't come soon enough." Someone was watching porn in the back row, recording it on his phone for later. People kept up their routines.

Outside, a protest against the lack of a real closure plan was kicking off. The Friends of the Library were joined by a handful of homeless patrons. A dozen people carried signs reading "We Are Human" or "What Would MLK Say?" They shouted for the TV cameras and sang together. Inside, twenty minutes before closure, staff were placing labels on every piece of furniture. They unplugged unoccupied computers and packed them away as patrons one seat over watched YouTube. Librarians took selfies at the desk together as the clock hit 5:30 p.m.

Homeless activist Cameron sang "Auld Lang Syne" in a soulful baritone. Police had no patience for ceremony, tapping seated patrons as they reminded them, "Excuse me, we're closing." People exiting joined the protest, now chanting "MLK! What would he say? Don't erase our public space." Cameron spoke with reporters, reiterating what he had told me: Everyone wants a renovation, but patrons also need services, and they don't need them on the city's edge. I walked by the Digital Commons the next morning, looking in through the glass walls that faced the Mather Studios condominiums. The computers were all packed up. The lab was empty. Four men were sleeping out front.

The Problem of Access

To close this investigation, I want to reflect on the power of the access doctrine to overrule these moments of resistance. Understanding this structural power may suggest new political strategies to combat it. Chapter 1 explored the access doctrine's birth in the neoliberal revolution, and the following chapters revealed how it is reproduced at the organizational level through bootstrapping. But the access doctrine clearly continues to play a larger structural role, enrolling people, resources, and organizations in an extensive project to redefine what the good life is and how to get it. It is a massive shift in the mode of social reproduction.

Making the problem of poverty a problem of technology changed how we understand the entire labor market, our navigation of it, and the reward from it—for employed, unemployed, and underemployed alike. Most accounts of the information economy focus on changes in the mode of production—the nature of work, what commodities are produced and where. And indeed, chapter 2, on startups and the perpetual pivot, is part of this genre. But what I hope I have shown through a subsequent focus on schools and libraries that see startups as a model is that there has been just as large a change in the mode of social reproduction, the differentiated process of making people for a particular political-economic moment. The access doctrine garners consent to this change, and the sacrifices it requires.

This new mode of social reproduction was not inevitable. It was man-made. Clinton (2000a) said, in his final State of the Union and the first of the new millennium, "We have built a new economy. And our economic revolution has been matched by a revival of the American spirit." The New Economy was built on services and software, measured in GDP growth. The "revival of the American spirit," the president said, was built on personal responsibility and measured in reduced crime, reduced teen pregnancy, and "welfare rolls cut in half to their lowest levels in 30 years."[1] Along with the New Economy, then, came new ways of acting on crime, children, the disabled, and the unemployed; and beyond these were changes, unmentioned in Clinton's speech, wrought by the neoliberal revolution in domains such as education, healthcare, and eldercare (Bezanson 2006). A shift in the mode of production required a concomitant shift in the mode of social reproduction. The access doctrine is an important part of this project. Understanding its role in social reproduction is an important first step in any attempt to act on that sphere of the political economy and, perhaps, rebuild it in a shape more conducive to human flourishing.

It is worth reviewing how we got here. This book explored how the problem of poverty is transformed into a problem of technology and how the organizations addressing the problem are themselves transformed in the process. This problem is relayed in hopeful terms as a *digital divide*: the gap between high-skilled professionals with access to personal computing resources and those without such access, a gap resolved by sending the resources and skills common on one side over to those on the other side so that the divided can become more like the connected. This access doctrine emerged from 1990s Clinton-era neoliberalism, when the persistent poverty visible within the New Economy was reframed as a lack of opportunities for competition. The access doctrine made it seem as though such opportunities were available to all via the internet, and thus any individual economic struggles were just that—and never the fault of deindustrialization, capital flight, stagnant wages, or a shrunken, punitive welfare state. This was part of the neoliberal revolution, wherein American political institutions redefined citizenship around market fitness and charged the state with either ensuring opportunities for competition and broad participation in them or policing those who could not or would not compete so as to preserve the smooth functioning of markets. The threat of punishment, of course, fell most heavily on the working and workless poor, particularly Black Americans.

This story about how poverty works is not absorbed via political pronouncement. It is reproduced daily as a means of understanding the problem and acting on it. Neoliberal urban development provides the stage for this project. It solves the problem of development through the digital divide and related projects like education reform, soliciting outside capital and outside talent to resolve the skills gaps present in the locals. The geography of the city is remade to support these moves, and institutions for keeping citizens alive and teaching them how to make a living are remade in the image of these largely White outsiders on the "right" side of the digital divide.

In startups, rapid personal and organizational change in an environment of economic uncertainty—the *pivot*—is essential for the internal and external legitimacy of the firm. In the library, this story prompts wide-scale reconsideration of the spirit and structure of the organization. Rebuilding the library as a training center for knowledge work earns political and financial support and makes the poverty, particularly homelessness, it is confronted with as one of the last public spaces in the city knowable and actionable. I called this process of constant reform in pursuit of the social mobility mission *bootstrapping*. We also saw it happen at W. E. B. Du Bois Public Charter High School. There, a conflict emerged. The school was dedicated to a particular set of social justice values, but its survival depended on demonstrating its success in human capital enhancement via high test scores and graduation rates. As graduation approached for the first senior class, the school's data infrastructure, ultimately controlled not by students or teachers but by administrators, was used to subvert those values and support a narrower, professionalizing vision of education.

These were the concrete institutional changes that came of reproducing the hope in personal computing and the internet. This bootstrapping had three primary causes. First, demographic similarities across largely White professionals grants faith in the power of skills training because it appears to work for them. Second, professional networks help train these workers in these organizational models and the ideas supporting them, and bridging organizations that connect these fields highlight startups as models and entrepreneurs as leaders. Finally, given the overwhelming uncertainty that comes from confronting a persistent problem like urban poverty, the comparative and highly visible success presented by startups provides a clear script for the organizations charged by the state with managing that poverty.

But schools and libraries cannot pivot like startups. They do not have the resources or flexibility. Much of their success or failure is beyond their control, and the bootstrapping project is itself constrained by an older, competing institutional culture based in public service—a resource for teachers, librarians, students, and patrons to drawn on in the hopes of preserving an ideal space for collective welfare. But people are being left behind, and so bootstrapping continues even as it fails on its own terms—often marginalizing the people those changes were meant to serve. Those homeless patrons, for example, who the library sought to empower through the Digital Commons are refused the safe space they need because sleeping does not fit the mission. The mission secures the organization, and so it must be preserved at all costs.

This organizational frenzy has its putative base in the problem of the digital divide, a technological instantiation of the skills gap narrative: technological advancement means American employers are short of the skilled workers—software developers in the narrow sense, STEM practitioners in broader narratives—they need, and so those on the fringes of the labor market can guarantee high wages and serve regional and national interests by training up. Learning to code serves not just your career, but regional and national economies. Skill matters for success in the labor market. But precisely what skill is and whether technological advancement (in the abstract or in the actually existing labor market) demands more or less of it is a matter of much debate. This is in part because, in practice, skill is often a proxy for workplace power: the ability of organized workers to claim their expertise, or the ability of employers to pay women and people of color less for the same tasks.[2]

The present study sidestepped the issue of whether these skills are actually in demand in the labor market and whether Du Bois or MLK are providing them: I simply have no data on the occupational successes or failures of my students or patrons in the years following their time in these places, nor would my ethnographic methods be suitable to answering the question of whether their learning experiences paid off in terms of higher salaries and greater job satisfaction relative to similar people who did not have those experiences. I cannot assess the success of the school or library on those terms. But, importantly, *neither can the vast majority of schools and libraries—and yet they still bootstrap in the hopes of producing these results*. More proximate measures, such as test scores, are adopted in part because they cannot observe their students' or patrons' labor market outcomes or change quickly enough to suit employer demands, demands that often differ and conflict across time, space, sector, and position in the business cycle.

The very fact that someone like the executive director of DCPL would fight hard to reframe libraries as "engines of human capital" (Martell 2016) shows that the meaning of skill is not a simple result of technological change. Rather, there is an ongoing, high-level political fight to redefine not just what skills are, but where and how you get them. To return to a question posed in the introduction, why would schools and libraries take on a training mission they cannot fully assess and for which they are not adequately prepared? A training mission that seems to exist in large part because employers, the parties who best know their own needs after all, are less and less interested in handling on-the-job training themselves (Cappelli 2015).

Making People Differently

Bootstrapping regularly fails on its own terms because the institutions it targets are not made for the task. And yet they still clamor to complete the task, in part because doing so bears fruit at an organizational level. This dynamic is a symptom of a new mode of social reproduction, wherein no one is ever truly on the outside of the labor market, but constantly searching for the skills, technologies, and opportunities that will help them move through it. This is a feature, not a bug, in neoliberal urban development. The access doctrine is reproduced by institutions of social reproduction—those spaces that do the work of making capitalist subjects—and it garners consent to these structural changes occurring within and beyond them.

This is insidious work because the ideal subject these institutions are redesigned to reproduce is an entrepreneur who has no need of these institutions; they can learn by themselves, work by themselves, start a tech company and weather extreme economic uncertainty by themselves. Institutions of social reproduction participate in this process because it provides short-term security. To chart a new path that resolves the crisis of care and builds power within institutions of social reproduction, helping professionals within these spaces will need to betray the class they share with tech professionals and struggle alongside patrons and students, rather than seeking to save them.

A New Mode of Social Reproduction

Whether private households or public schools, the places that make people occupy a contradictory space within capitalism (Bhattacharya 2020). At a local level, their work need not turn a profit and so they can be refuges from the violence of market competition. On the other hand, their work is violently conditioned by those same market imperatives. Households run on hard-won wages, schools and libraries on taxes that must be fought for— and the women's work that largely keeps both spaces running is ruthlessly devalued, to zero in the case of household labor that is not commodified in the form of nannies, cleaners, and so on.

The inputs to households and institutions also emerge from the market and are subject to its vagaries. The price and quality of food varies like any commodity, but you need dinner to work the next day. A day care needs supplies for its children, and if it cannot afford them, then teachers may well pay for them themselves. Outputs are similarly conditioned. These spaces must produce not people in general but workers under capitalism specifically—and a whole host of attitudes, values, and behaviors are required of those people if they hope to survive.

At a structural level, capital requires socially reproductive labor to maintain its own circulation. Capital cannot make people, but its need for more and different labor power is balanced by an impulse to disinvest from the costs of social reproduction and disrupt or abandon the spaces of solidarity and community that are grown therein (Katz 2001). Fraser (2016) argues that this latter dynamic creates a "crisis of care" in each capitalist epoch. Our current crisis is marked by the state's retreat from the responsibilities of care and the stagnation of wages for much of the working class, leading to a "dualized organization of social reproduction, commodified for those who can pay for it, privatized for those who cannot—all glossed by the even more modern ideal of the 'two-earner family.'"

I would add that the state has not so much retreated from care but shifted those capacities into punishment (Gilmore 2007)—particularly for the racialized working class—and that this is complemented by capital's retreat from providing training within the firm. Capital has shifted the burden of skills training onto individuals so that they must now have all the skills necessary to succeed at jobs they do not yet have. This shifting of the burden of risk and the duty of care from the state and corporations and onto individuals also appears in other domains, such as retirement planning (Hacker 2019). To carry the burden of skills training, individuals take on the burden of debt, access to which is also racially differentiated, in order to pay for college or coding boot camps (Cottom 2017; Seamster and Charron-Chénier 2017). If they are on the fringes of the labor market, they are forced to enter a constant cycle of job search, job applications, and skills training; this is the replacement of welfare with workfare (Peck 2001). Police and prisons handle those individuals who will not or cannot shoulder these burdens.

For the institutions that make people, this burden shift looks somewhat different. Particularly during the postrecession years that were the focus of this study, the state retreated from consistent funding of social reproduction. The available tax base shrunk. Even outside of recessions it is now difficult for municipal politicians to justify tax hikes that they feel might weaken their position in an environment of intensive inter-regional competition, what Harvey (1989) called *entrepreneurial urbanism*. But, as we have seen, in this moment of fiscal retreat and despite their own weakened resource base, places like schools and libraries jump into the breach. These organizations are eager to take on the skills-training burden and assist individuals with the risk shift because they are themselves at risk and they hope this will provide some security.

The ideal economic model for individuals at this juncture is the tech entrepreneur, just as the ideal model for organizations is the tech startup. "Every day is a school day" was the mantra at InCrowd during my time there. The ideal entrepreneur is a "roaming autodidact" (Cottom 2016) who seeks out the unlimited educational opportunities available on the internet. They are self-sufficient, not reliant on institutions of social reproduction. The myth of the self-taught, college-dropout billionaire is a powerful one, though it runs against actual data about the importance of educational credentials in the tech sector (Weber 2016). The ideal entrepreneur keeps training throughout the life course. It is as an investment in their own future, where the career direction and location might change at any moment (Neff 2012). The ideal entrepreneur is a young White male who not only lacks any responsibility to care for others, but also has his own care needs commodified, provided either by an on-demand delivery app or by his workplace.

The contemporary shift in the mode of social reproduction is based on making more of these ideal entrepreneurs, not necessarily because we need them but because they supposedly thrive in an environment of privatized care and general macroeconomic uncertainty. The access doctrine secures consent to this crisis of care and this atmosphere of uncertainty. By turning the problem of poverty into a problem of technology, economic security becomes a matter of getting the right tools and the right skills. Individuals must seek out these resources themselves, and institutions can only survive by assisting them.

While these messages dominate our popular culture, consent is not secured through propaganda. It should be apparent throughout this ethnography that no one involved in bootstrapping their organization was tricked, nor were they attempting to trick the people they served. The function of the access doctrine, then, is not just in its ability to persuade the masses but in its ability to mobilize different classes and different institutions into a new hegemonic bloc. Hegemony, Gramsci reminds us, is the substitution of one class's interest for the rest of a society's, secured through alliances with and concessions to other classes. Ford's five-dollars-a-day wage is the classic example here: high for its time, but necessary to guarantee participation in the violent grind of the assembly line.[3] This helps to explain why bootstrapping institutions are so eager to take part in a project that so often does not meet the real needs of the people they serve, if it does not directly harm them, and which, over the long term, supports an entrepreneurial vision of economic life in which they themselves are unnecessary.

This project secures a place for schools and libraries in contemporary capitalism, even if that means an alliance with the politicians and entrepreneurs who previously called these places outdated and unnecessary. This is why DC Public Library shared the stage with the DC Chamber of Commerce at the digital divide event in chapter 1, or why startup leaders were on center stage for the Education Innovation Summit in chapter 5. It is the high-level version of why MLK's illustrations of its future included a picture of a TED Talk and why it practiced that future by kicking out sleepers. The state plays a leading role in coordinating this project, assuming that what is best for tech is best for the city. City leaders wagered that an influx of wealthy White migrants would uncouple DC's economic fortunes from those of the federal government.

Besides changing the composition of DC's workforce, tech's hegemony and all those new (mostly White) people support residential and commercial real estate development. Part of the brilliance of the creative class development strategy (Florida 2004) comes from the fact that creative workers in startups require very little fixed capital, so introducing housing and services that cater to this class is framed as an investment in the sector's productive capacity. When the class you are building a city for can work anywhere with their laptops, the normal violence of gentrification—the replacement of affordable housing with luxury condominiums, of local businesses with expensive national chains—can be understood as an industrial development strategy.

Not only were DC's startups concentrated in increasingly expensive, gentrifying areas, but the city events promoting DC Tech were often hosted in new spaces looking for tenants. This might look like Vincent Gray's Digital DC Thank You at the O Street Condominiums, just about to open, or the Entrepalooza startup fair, housed in a massive, vacant commercial space, a Borders bookstore before the chain went bankrupt. This is the work it takes to make "The Internet: Your Future Depends on It" hegemonic.

Actually Existing Ideal Workers

But of course, every institution has a human face. Startups, libraries, and schools are nothing without techies, librarians, and teachers. The hegemonic bloc that implements the access doctrine is formed in part by an alliance between these groups, who appear to form a class at the center of new modes of production and social reproduction. Throughout this study, I have remarked on the similarities among these different groups. They were often White newcomers to the city, with similar workstyles and a professional commitment to their craft. Uncertainty pervaded their work lives, but there was a meritocratic faith in their ability to join with their skilled colleagues and overcome it. They represented the leading edge of a new city. From a thousand-foot view, this looks like a unified set of White professionals remaking DC in their image—and remaking the people in it as well; recall student Irene's observation that Du Bois was "so good at trying to prep us to be like preppy White folk."

But there are important differences within this class. Up until now, I have largely used *digital professionals* to describe these people and what they do, designating the subset of teachers and librarians *helping professionals*. A class, however, is not a static set of demographics or tasks, but a historically specific set of relationships with both capital and other classes (Gunn 1987; Thompson 1991).The terms "digital professional" is sufficient as a description of an ideal type: a self-directed employee or entrepreneur whose work tools are information technology and whose work tasks are the production and circulation of immaterial commodities (e.g., software, blueprints, financial products; see, e.g., Drucker 1999; Hardt 2005). It is insufficient for the purposes of describing an actually existing class and the schisms within it, much less its relationship to other classes and the broader mode of production. An older idea, that of the professional-managerial class (Ehrenreich and Ehrenreich 1977, 2013), helps fill the gap.

The idea of the middle class has always been a powerful one in the United States, in part because it is so underdefined as to include everyone but the poorest and richest Americans. But the Ehrenreichs observed a very real class in formation throughout the twentieth century, beginning in the progressive era and taking a leftist turn in the campus revolts of the 1960s. Over the course of the nineteenth century, it had become clear that the work of managing people, information, and technology within complex, globe-spanning firms could no longer be left to individual entrepreneurs. Similarly, from capital's perspective, the multiethnic, multiracial mass of proletarians living and working together in US cities at the turn of the twentieth century demanded a new set of professionals who could enforce broad standards of health, education, and American culture—work often funded in the progressive era by both the state and those same industrial capitalists (e.g., Carnegie, Rockefeller). People would not become workers on their own.

This is the work of the professional-managerial class: reproducing capitalist culture and inter- and intra-class relations. That work happens either inside firms, through direct management or the implementation of technologies and techniques that maintain the shape of the firm, or outside of firms, through the work of social reproduction. One half of this class, represented by InCrowd, provides products and services. The other half, represented by MLK and Du Bois, provides care. This class does not own the means of production and so has an antagonistic relationship with the capitalists who control its livelihood; but it also has an antagonistic relationship with the working class it molds to fit the mode of production. Think of the tension not just between shift manager or human resources and a regular employee, but the relationship between social workers and their clients. Frequently, these relationships are racialized, such that Whites, especially Protestants from Northern and Central Europe in the progressive era, were managing ethnic Whites and non-Whites. This class depends upon education to advance and frequently organizes not in unions but via professional associations.

Observing this class's ascent to political power in 1983, Darity noted that the new US political economy required not just the substitution of heavy industry for services, but new institutions of governance—some educative, some carceral—that would clarify the relationship between those at the center of the economy and those at the margins. He predicted that the working and workless poor, disproportionately Black, would need to be trained or contained, with ever more creative solutions revealing the unchanging fact of persistent poverty in the information economy: "If the managerial society finds any use for its outcasts or undesirables, it will be as objects of experiment. On those who do not matter, there is a willingness to try anything" (59). Both Du Bois and MLK testify to this ongoing experimentation in the face of structural marginalization. As did librarian Grant's admission that he often felt there was little they could offer homeless patrons besides entertainment. This shows that that training and containment, the school and the prison, are not binary choices for managing surplus populations, but an institutional continuum through which the professional-managerial class, either in their jobs or in the policies that emerge from their coalitions, sort through the real people who make up the stubborn fact of persistent, racialized poverty. Where neoconservatives expressed this in starkly violent terms—wars on crime or drugs—the innovation of the Atari Democrats was to offer the access doctrine as a hopeful story that could take the working and workless poor from one end of the continuum to the other.

This political position is in no way inevitable. Owing to their liminal class location, the professional-managerial class has the potential to split with the capitalists. This may happen by choice, as with the campus revolts, environmental movements, and antiwar demonstrations of the 1960s.[4] This may happen because of an assault by capital; deskilling within the firm and austerity outside it quickly removes the privileges of professionalism. Since the 2008 recession, corporate deskilling and state austerity have engendered a great deal of downward mobility in the professional-managerial class. These status-threatened professionals and their déclassé children have powered such movements as Occupy and the post-2016 boom in the Democratic Socialists of America (Press 2019; Winant 2019). But the growth of such a movement is never guaranteed, much less its success, and hard limits are placed on those that do emerge because the structural function of this class is to reproduce the social relationships required by capitalism, and this function grants real advantages over other workers.

The persistence of the access doctrine, the organizational changes it wreaks through bootstrapping, and its role in neoliberal urban development is secured through an alliance between the two sections of the professional-managerial class: those reproducing capitalist relations within the firm (tech workers at InCrowd) and those doing the work of social reproduction outside the market (helping professionals at MLK and Du Bois). Capitalists cannot simply speak hegemony into being. Remaking a city in tech's image requires both an influx of new tech workers from the outside (InCrowd workers, in this book), and the transformation of the existing populace in support of this new project (the teachers and librarians). Startup workers pursued this project inside their firms and, through their products, for other firms. InCrowd's engineers, perhaps more so than their coworkers in other units, were conscious of their class position because they built the software caterers used. Paul noted in one team meeting, immediately after the company as a whole had met to hear updates about its financials, that "we are in the business of firing people [specifically food service staff], that's what 'we help your company be more efficient' means." A much more hopeful evangelism suffused DC tech, especially in incubators like 1776, in order to make a place for the sector in the city.

As we saw, the helping professionals at Du Bois and MLK were more ambiguous about their class position. Unlike techies, their work required regular interaction with the city's working and workless poor. Like techies, they were mission driven—both because the survival of their organizations depended on enthusiastically bootstrapping and because they were themselves committed to a public service role. And that public service role was increasingly under attack: recall Amanda's exhaustion and the Du Bois employees who quit mid-year, or the stress librarians felt in trying to fulfill tasks better left to a social worker when none were forthcoming. The city built in tech's image places great strain on their workplaces. The tools built by tech (e.g., SchoolForce) restructure their jobs, undercutting the autonomy that draws people to these professions in the name of greater transparency.

New threats come clothed in the hope the access doctrine invests in their mission. In the summer of 2020, DCPL proudly shared praise of the new MLK from the *Washington Post*'s architecture critic. Under a headline reading "America's libraries are essential now—and this beautifully renovated one in Washington gives us hope," he extolled the light, airy building's potential to fulfill the promises the welfare state made to its citizens, especially those Black Americans excluded at the welfare state's genesis (Kennicott 2020). But at that time, Eugene, now in a leadership position in the librarians' union, was negotiating with management to preserve his profession's place in that hopeful future. He was not optimistic: "The deskilling of first-floor staff is pretty close to complete; there will not be any librarians or library associates working in non-supervisory jobs on the first floor." Despite sharing a lifestyle and a race-class position, there is, then, intraclass antagonism between the helpers and the tech workers. In the sphere of social reproduction, their institutions mandate an interclass antagonism between these professionals and the people they purportedly serve. Homeless patrons with nowhere else to go were dismissed from the library for sleeping. Black and Latinx students were molded into a particular shape. There were also individual moments of compassion and camaraderie that strayed from this bootstrapping script, often based on an older public service institutional culture. Librarian Grant did everything he could for "computer man" Shawn. Clara taught her students to do countergentrification action research and helped with logistics for their Black Lives Matter walkout. But these moments are not a movement that will undo the common sense of the access doctrine.

Such a movement will require a broad, organized assault on the neoliberal political apparatus. Had I the recipe for such a program, I would gladly share it—but that is beyond both my ken and the remit of this book. However, I am confident that our cities can be if not remade, at least equalized, and the bootstrapping cycle broken by the people within these institutions. It will require a split within the professional-managerial class. Helping professionals must break against the tech workers with whom they shared college dorms and now share apartment buildings. Instead they must organize at the point of reproduction in solidarity with their patrons and students, based on the recognition that both helper and helped share an interest in preserving the institutions of public service threatened by the access doctrine.

Strategically, both sides need each other. The helping professionals are few in number, and state and capital can easily paint their protests as dereliction of sacred duties (e.g., "strikes hurt kids"). They need support from the communities they serve in order to stand up to this political pressure. On the other hand, students, patrons, and their communities have numbers but lack the strategic position within the mode of reproduction that helpers' work provides. A protest in support of library patrons is one thing, but a library shut down by its workers is another thing entirely—revealing just how much the city relies on these institutions to function and the sort of power these professionals have. This work is already happening across the country.

Organizing for Hope

The protest over MLK's closing was too little, too late. Homeless patrons like Cameron had a tremendous amount of political literacy but were, collectively, demobilized by the daily assaults of urban poverty. They did not have a long-standing relationship with the Friends of the Library—the "nice White people" Grant felt he had to shield from patrons—and while the Friends worked hard, they did not occupy any sort of strategic position within the library that could challenge its decisions. Librarians largely did not participate because they had a job to do.

Centers of social reproduction like those studied here are not victims of neoliberal reform but sites of struggle in which countermovements can be built. Schools and libraries and places like them bring together in one building a professional-managerial class under pressure to meet their metrics, students and patrons competing for their own futures, and parents and community members who rely on these sites for childcare, food, bureaucratic assistance, adult education, and, of course, free Wi-Fi. Workplace conflicts here intersect with broader community conflicts over gentrification, policing, and the provision of public resources. Workers in these institutions are, like many helping professionals, under assault. But unlike those members of the professional-managerial class who are isolated from the working class they manage, teachers, librarians, and the like are, in the course of their everyday duties, directly confronted with the dual assault on both their own organizations and the livelihoods of the people they serve. In this book, that class consciousness has led to, at best, ambiguity. But other schools, libraries, and cities are possible.

The crisis of care Fraser (2016) identifies has recently prompted an organizing wave in social reproduction sectors, where the rallying cry is often a variation on that of the Chicago Teachers Union (CTU): "Our working conditions are students' learning conditions" (Uetricht 2014). This political philosophy unites the professional-managerial class running the school with students and neighbors, and it has ensured that the string of teachers' strikes that began with the CTU and has since extended across the United States has enjoyed broad popular support (Blanchard 2018). These struggles make clear that the push for, for example, high-stakes testing diminishes not only educators' autonomy, but that of their students. Teachers in Los Angeles went on strike not only because of budget cuts that were crowding their classrooms or curtailing their health insurance, but for an end to random searches of students of color by school security officers (Henwood 2019).

The power of these strikes lays not only in the interracial, interclass alliances, but in how they lay bare what Katz (1998) calls the "hidden city of social reproduction": the care of teaching, helping, nursing, and so on that makes capital accumulation possible, but which is at best neglected by capital and at worst sabotaged by it. As United Teachers of Los Angeles President Alex Caputo-Pearl put it, "The reason they're driving privatization so hard is because we know they can't offshore. They can't move the hospital to China. They can't move the school to Myanmar or to China, so instead, the neoliberals have set out to decimate the teachers' unions" (Henwood 2019). The neoliberal assault on these institutions underlines their strategic importance.

Schools and hospitals are perhaps the most advanced sites of struggle in this framework. But librarians have organized with and for their patrons to demand better and could well do so again. In the late 1960s and early 1970s, as public sector unionism grew in the United States, a series of strike actions gripped public libraries across the country as librarians fought for better wages and hours but also expanded services and control over them (Chaplan 1976; see also McCartin 2006). A picture from the picket line of a 1968 strike by Contra Costa County librarians in California shows three bespectacled White women sternly regarding the camera; a placard on the middle woman's lap reads, "My loyalties are to PEOPLE, not institutions" (AFSCME 1968). In 2016, Emily Drabinski, a librarian at Long Island

University, led the university's faculty and staff through a lockout but only emerged victorious because students organized in solidarity. "They've tried to put a wall between us and the students, but it hasn't worked, because it's hard to be on anybody's side but ours in this," she said. "They fucked with the wrong students" (Bonhomme 2016).

This sort of social justice unionism faces incredible barriers. Organizing with the community in which you are based is a strength, but it requires negotiating the unequal power relations that structure interactions between the often White, female, and middle-class helping professionals and their often poor or working-class patrons and students of color. Recall for example the sticker system Rachel developed to privately divide gold-star patrons using the library correctly from those who were not, those she called "dumb as paint." Or Sam's uneasiness with the White savior mission he saw in TFA. These divisions become common sense through a combination of racialized professional identity and organizational strain. A great deal of work is required to overcome them.

Even empathy with those on the other side of the digital divide is insufficient. The helping professions base their daily reproduction of capitalist relations in personal benevolence for those on the other side of various racial economic divides. Ceding control to the divided, organizing together to run the institutions democratically, is something else entirely. But this is what is necessary to stop the endless training treadmill and build spaces that build power for threatened professionals and the working and workless poor alike. What could our schools and libraries do if they did not have to try to be startups? Could DCPL offer credit union services in neighborhoods without retail banking? What if it really was a safe space for people to sleep? Could Du Bois add to its weekend test-prep and college application programs further adult education services for parents learning English? Could these progressive schools become bulwarks against police violence instead of teaching students their rights on one day and searching students like Martin for pot on another?

The seeds for something different are there. Amanda was the OWL at Du Bois because students treasured her classroom as a safe space; they would learn to code in class but could talk about their home lives after class. Homeless activist Cameron was especially disappointed in how the city handled MLK's closure because he saw so much promise within the library, a different vision of solidarity. The late Pamela Stovall, former associate director of

DCPL and subordinate to Reyes-Gavilan's predecessor Ginnie Cooper, was a powerful figure for him in this regard. Years earlier, she had joined Cameron and his colleagues in a prayer circle at the old First Congregational building. They told her they felt abandoned by their city. So Stovall kick-started a program at MLK called "Your Story Has a Home Here," where teen-aged patrons were trained to conduct oral history interviews with homeless patrons. Photos and excerpts were displayed throughout the library, and some are still accessible online. In "Your Story Has a Home Here," Cameron saw solidarity and recognition.

When Shawn told me "I was always a computer man," he was emphasiz-ing exactly this sort of recognition, of the creation of community bonds that do not necessarily fit the helper-helped dichotomy. "I was always a computer man" is a signal that he could take what he needed from both the ideal bootstrapping library that provided those computers and the ideal public service library that provided a refuge, overlapping as they did in the actually existing library building, and craft a secure space for himself amid the rest of his life's uncertainty.

He and Ebony worked hard to find that security in computer labs all over the city and were pushed out of some of them by library police. But in MLK they found a space they could fit to their needs and friends who supported them in it. Sometimes staff, like Grant, supported them too, but the bootstrapping project narrowed the sort of assistance they could pro-vide and overrode individual voluntarism. For Shawn's political vision—a computer lab that is always secure, always supportive, even if you're unpro-ductive—to be generalized, collective action at the point of reproduction is necessary. The access doctrine makes libraries that do not support this insecure. So they must be remade.

The protests by the Friends of the Library and Cameron and his col-leagues over the library's closure lacked the sort of force that would make the city notice. The sort of solidarity necessary to create the alliances that power truly disruptive actions takes a long time to build. Many librarians and homeless patrons were suspicious of the Friends. Librarians had worked hard to direct patrons to resources they could seek after the closure, but the momentum of the new library had built very literal divisions—like the glass walls of the Dream Lab that guarded renovation planning meetings—between the two groups. The work librarians had to do to keep the library going was not the work many wanted to do to support the homeless patrons they saw all day, every day. But recent strike actions show that this need not be the case. Solidarity can be built, and it can win.

This vision of organizing the community at the point of reproduction may, in contrast to the preceding chapters, seem downright antitechnological. But particular implementations of information technology have started these fights, and different implementations could win them. The library computer queue or the SchoolForce student data system had particular values about how the institution works built into them, reproducing the social relations of the bootstrapping space even before any librarian or teacher intervened (Winner 1980). Similar technological projects have kicked off the social reproduction strike wave reviewed earlier. Notably, one of the last straws for West Virginia teachers was a proposed requirement that they wear Fitbit devices at work in order to collect movement data that would determine the cost of their health insurance premiums (Boothe 2018).

Different visions of how information technology can work and for whom are already visible. A whole new set of digital literacies were needed to connect teenagers, librarians, and the homeless in the "Your Story Has a Home Here" project. However, it must always be remembered that these are not just tools, amenable to any use. Reversing or reworking the logic of bootstrapping will inevitably require restructuring both the technological infrastructure undergirding these organizations and the institutional culture driving their curricula.

This must be a democratic project because, as we saw at both MLK and Du Bois, control over these systems is not evenly distributed. But I believe the seeds within these places can, with care and struggle, expand and transform the space around them. For those helping professionals interested in working with their community to build technologies that build power, it is worth remembering that techniques of participatory design were not always the broad rubric we know today: focus grouping and codesigning new tools for particular contexts to best serve particular constituents. One origin story for this design methodology comes from efforts to build workplace power: Scandinavian trade unionists working together to increase their understanding of workplace technology relative to their managers' and intervene in new technological implementations before they were set in stone (Asaro 2000; Beck 2002). The experience of redesigning institutional technologies can be a space of solidarity. How could, for example, the queue at MLK be redesigned to reflect how long it actually takes to fill out a job application? How could teachers and students work together to rebuild SchoolForce in a way that supports Du Bois's racial justice values, rather than its metric-chasing mission?

At a structural level, the drive to bootstrap comes in large part from the limited options neoliberal urban development lays on the table. Teachers and librarians empathized with the people they served and hoped for all sorts of solutions to their problems. But human capital enhancement was what was on offer, in large part because that kept the organization alive. This recalls earlier moments of political struggle in DC and similar cities. Urban politicians, largely Black, sought out a variety of political solutions to the crime wave that began in the 1960s and subsided in the 1990s. But social programs that would assuage the social dislocations behind crime were too expensive for cash-strapped cities and unsupported by state and federal politicians interested only in punishment. And so police and prisons were the solutions chosen (Forman 2017).

This cycle need not repeat. Teachers organizing with their students and neighbors in cities like Chicago and Los Angeles show us that budgets can change, new services can be added, and new institutional scripts written. A new kind of democratic public service organization can be built for a more just economy, and the failed cycles of bootstrapping can be left behind.

That which emerged from struggle can be changed by it. If nothing else, I hope I've shown that the story driving so much of our thinking about poverty—"The Internet: Your Future Depends on It"—did not appear out of thin air. It had to be told over and over, reinforced through web filters, progress reports, and planning documents. We built these coping strategies to make overwhelming economic inequality sensible and navigable. But if access today means, fundamentally, an opportunity to compete, then an alternative should not be so hard to imagine. Because if the world as it currently exists is one where we must be granted the tools necessary to strive for excellence, to innovate beyond our current dire straits, to outcompete inequality, then surely another world is possible where innovation is boring and excellence is unnecessary because the good life is ordinary. What would we compete for if so many would not starve for losing? There is so much work we have to do.

Notes

Introduction

1. Any use of a first name is indicative of a pseudonym. Full names indicate that the person being discussed is a public figure and that any quotes from or descriptions of them are drawn from public events. The lone exception is Principal Carroll, a pseudonym that required a position.

2. During the period of fieldwork, 2012–2015, the percentage of US adults who reported using the internet regularly plateaued at 84 percent. Class-related gaps shrunk between 2000 and 2015, but households earning more than $70,000 annually are still more likely to be online than households earning $30,000 or less. Similarly, racial gaps exist but have decreased: 78 percent of Black Americans and 81 percent of Hispanics use the internet, compared to 85 percent of Whites and 97 percent of Asian Americans (Perrin and Duggan 2015). These measurements do not account for the quality of service or inequalities that exist in use or the rewards for use.

3. This too is complicated. Although productivity has continued to increase since the 1970s, the rate of growth has not come close to matching that of the postwar Keynesian golden age and has nearly stalled in the twenty-first century. This another instance in which it becomes clear that the "golden age," the gains of which were of course unevenly divided by national and international expressions of racism, was an exception rather than the norm, and that the economic imaginaries built within it simply do not fit our present circumstances (Gordon 2016; see also Baily and Montalbano 2016; Benanav 2014).

4. In this vein, I draw inspiration from *Policing the Crisis* (Hall et al. 2013), which demonstrated how an early 1970s English moral panic over a spike in muggings—vastly out of proportion to, and indeed actually preceding, any real rise in rates of street crime—was produced by the national government and reproduced by police, the judiciary, and the mass media as a way to manage social tensions produced by the collapse of the post-WWII economic boom and the welfare state that matured within it. Those tensions were pinned on a scapegoat: the urban Black youth who

had been largely left out of that boom and were suffering most from the crisis of the 1970s.

5. Bootstrapping, as a process of organizational restructuring meant to fight poverty through technology, evokes both the American mythology of the self-made man and the technical process of restarting a computing system and loading successively more complex programs from a simple starting point.

6. See "Greed City," chapter 8 in Sherwood and Jaffe 2014, for an account of the 1980s and 1990s commercial property boom in DC. The city was a major innovator in tax increment financing (TIF), using it for place-building projects such as Gallery Place, the Mandarin Oriental Hotel, the City Market at O Street, and the DC-USA mall. TIF pays for contemporary economic development initiatives through future, earmarked tax revenues produced in that area or by that development. Across the United States, it has become a key financial mechanism for neoliberal urban development. See Weber 2002 for a fuller description of this funding mechanism.

7. For criticisms of the role of hope in urban development not directly related to technology or the tech sector, see Anderson and Holden 2008 and Wolch 2007. Anderson and Holden describe hope as an affective infrastructure used to enroll Liverpudlians in their city's urban renewal project, centered on the city's 2008 designation as the European Capital of Culture. Wolch's study of the Los Angeles River is an intervention into the climate of fear that dominates discussions of urban ecology, showing instead how human-nature collaborations can nurture a different kind of noncompetitive urban citizenship, built for the long term.

8. Following other ethnographic methodologists, such as Calarco (2018) and Glaeser (2005), I see the value in this inductive research method in its ability not to, as statistical methods do, generalize to a particular population, but instead to generalize to a particular social process. This is one reason that ethnographic contexts also become ethnographic genres: workplace ethnographies, school ethnographies, and so on. The other, less scientific advantage to this method is that done well, it tells good stories that can illuminate human affairs in a way numbers simply do not.

9. Unlike MLK, both InCrowd (and the other startup featured in chapter 2, Hearth) and Du Bois are pseudonyms. Minor details about the startups' business models and the school's location and appearance have been changed to support their anonymity.

10. I describe the two processes—pivoting within startups and bootstrapping within schools and libraries—with different names because, despite public service organizations' attempts at mimicry, they have different motivations and different results owing to their different resources, personnel, and purposes. MLK and Du Bois might wish to remake themselves in the image of the startup I call InCrowd, for the sake of their own security, but they cannot.

11. Far from a disciplinary figure, I actually built rapport through my accidental habit of breaking the spaces' rules—until I wised up. Like patrons, I brought food

into the computer lab because I didn't want to lose my seat. Like students, I broke the dress code several times because I really did like my new sneakers.

1   Discovering the Divide

1. Many thanks to Nathan Ensmenger for referring me to these articles.

2. If the amount of monetary aid any given state is able to distribute annually is capped by the federal government in advance, states have a great deal of difficulty in responding to periods of increased joblessness—when requests for aid skyrocket.

3. Throughout this chapter, I use the term *New Economy* to capture the shift in the mode of production summarized in the introduction, but only because that is the term the Clinton administration used in its own hopeful economic policy documents. Obviously, two decades into the twenty-first century, this state of affairs is no longer new in any empirical sense, though arguments for its novelty still retain significant rhetorical power. And so I use *information economy* throughout the rest of the book, to capture the institutional drive to remake organizations and people for an economy based primarily in the production and circulation of information. These are emic terms, derived from the groups under examination rather than an endorsement of the idea that the labor most socially necessary to the reproduction of contemporary capitalism is in software development. As the introduction, particularly the BLS projections, makes clear, it is not. In the United States in particular, contemporary capitalism is dominated by low-wage service work in food, hospitality, and healthcare.

4. The final year in which the FCC reported these data was 2016. The Trump administration's FCC did not report data on market concentration in this sector and argued that penetration should always be a composite measure of mobile plus fixed terrestrial broadband, thereby obscuring the limits of broadband adoption and its oligopolies.

5. Some of that is made up for by our comparatively higher ranking—fourth—in mobile broadband subscriptions, but mobile-only users have a much harder time with essential, typing-intensive tasks like searching for jobs or completing homework (Smith 2015).

2   The Pivot and the Trouble with "Tech"

1. InCrowd and Hearth are both pseudonyms. Elements of the startups' business strategies and branding have been altered to preserve anonymity, while still retaining the core features of their revenue models, operations, and identities.

2. *Bridging organizations* was originally a term used in the development literature (Brown 1991) to describe those nongovernmental organizations that link scattered antipoverty initiatives and their constituencies into a broader movement. I use it to describe those organizations devoted to sharing resources and ideas among similar

organizations and advancing the cause of a sector more generally (Reveley and Ville 2010). In the private sector, this may mean voluntary industry associations in which business leaders participate, professional associations in which workers participate, regulatory agencies that promote local development of a field or define its credentials, or training organizations that bring newcomers into the field. In the public sector, we may add to this list those philanthropies that fund new initiatives in the field, convene discussions on the field, and thus shape its priorities. It is important too to recognize the role that private sector bridging organizations play in remolding the public sector.

3. Two persistent myths obscure the division of emotional labor in startups. First, the very real sexism in the industry and the declining presence of women in computer engineering and computer science majors since the 1980s (NSF 2013) leads to the assumption that women are largely absent from the tech sector or at least inessential to it (e.g., Khazan 2015). Women may be undervalued in tech, but exploitation is not absence. It is, rather, a sign of the many types of paid and unpaid labor that are crucial to the maintenance of the company but are not paid fairly because this work is perceived to be either unskilled, in comparison to software development, or the sort of caring, interpersonal work women are supposed to take to naturally. As Kate Losse's (2012) history of Facebook's early years showed, women in startups find themselves playing a variety of roles beyond those listed in their official duties (e.g., planning events, mediating disputes), many of which are crucial to the continued success of the firm (e.g., Losse's work as Zuckerberg's ghost writer) but do not match the cultural ideal of the young, White, male software developer. Second, the very real focus in the very competitive industry on quick-moving business plans and computer science skills leads to the assumption that tech does not care about or think deeply about culture. Of course, everyone has culture and all organizations build a specific—if sometimes implicit—internal cultural in order to keep themselves going. Startup engineers may have a very different understanding of culture from humanists, but they still research it, act on it, and actively promote it in their organizations (Seaver 2015).

4. Office tours were a fixture of the DC startup scene—so much so that DC Inno, the local industry news portal, ran a regular Office Envy column with breathless photo tours of startup offices.

5. Five years later, WeWork would become the face of tech excess. Multibillion dollar losses and no clear path to profitability meant its business model—subleasing office space to entrepreneurs—collapsed and its IPO unraveled (Platt and Edgecliffe-Johnson 2020).

6. Since the 1990s, DC has been notoriously bad at maintaining its jobs programs. The Department of Labor ranked it dead last among "states" in terms of effectively using federal job training funds. Since 2012, Labor has classified DC as a high-risk partner for job-training programs (McCartney 2015).

7. DC is effectively a one-party city, so the Democratic primary in spring is the de facto general election for mayor.

3 "More Than Just a Building to Sit In for a Day"

1. The term *homeless* is contentious. Homeless activists, nonprofits fighting housing deprivation, and critical scholars of housing have offered up a variety of alternatives. Some emphasize both the people involved and the fact that homelessness is a social status rather than a permanent identity (e.g., "people currently experiencing homelessness"). Others shift the focus to the capitalist housing system that creates homelessness in the first place (e.g., *houselessness* or *the unhoused*). *Homeless* was the term most often used by my informants, whether homeless or not. I use it for that reason and because, following Wilse 2015, I am exploring homelessness as a particular category of people constructed as a problem by particular state agencies, nonprofits, and private developers concerned not so much with solving housing insecurity but with managing the problem-people in a way that not only minimizes their disruptions to urban development but that also becomes economically and politically productive for the problem-managers.

2. Library fieldwork focused on MLK, supplemented by special events at and occasional visits to other DCPL branches. Library interviews drew from patrons who visited MLK frequently and from librarians who worked both at the central branch and satellite branches—in part to better understand whether the organizational culture at MLK generalized to the rest of DCPL. *Librarians* in this chapter is a broad term encompassing all library workers who work with or for patrons and handle informational materials, thus excluding custodial workers or police but collapsing the distinction between entry-level library technicians and those with master's degrees in library science.

3. Mecanoo is a Dutch firm. Martinez + Johnson were local, and had redesigned DCPL branches in Takoma Park and Georgetown. Over the course of MLK's renovation, Martinez + Johnson were acquired by a larger, national firm, OTJ Architects, who continued their work.

4. The Advisory Neighborhood Commissions (ANCs) are the smallest unit of local governance in DC. Each ward is divided into between four and six ANCs, and each ANC has a significant say in what development is approved or blocked in their neighborhood. They are thus important power brokers with which even large developers or municipal officials like Reyes-Gavilan must curry favor. In theory, the ANC structure allows for significant community input into local development. In practice, the people speaking for the community are rarely representative of it: typically older, wealthier, Whiter homeowners with the literacy and free time available to provide input to these meetings. Williams (1988) discussed this process as part of her investigation into an earlier wave of DC gentrification. For a more recent review of similar neighborhood governance dynamics in Boston, see Tissot 2015.

5. The 2000 Children's Internet Protection Act requires all libraries receiving federal funding to install such filters.

6. Patrons who were relatively new to the library and in need of a quick transaction (e.g., tourists printing out a ticket for the nearby Chinatown buses that go up and down the east coast, date-night couples checking the Regal Gallery Place movie theater schedule) would often walk into the Digital Commons, sit down at an empty PC, try to log on, and stand up flustered before being redirected to the sign-up queue. They were inevitably younger and Whiter than regulars.

7. Interestingly, the word *science* does not appear in the college's name, although it does of course appear in the MLS—and the iSchool itself, now my employer, grew to include more "scientific" degree programs such as master's degrees in human computer interaction and information management and a BS in information science. And though the MLS curriculum in general did shift to include more of a research base, Becca does appear to be projecting her fears over the profession's future onto the college as a whole.

8. In many ways, this is a return to form. As Kernochan (2016) notes, the recent turn to foundation funding echoes the early days of US public libraries and their reliance on Gilded Age industrial magnates—most notably Andrew Carnegie—although this time around the funding must be sought through grant competitions.

9. For further discussion of the default whiteness and default professionalism of digital spaces and creative misuse of them by non-White technologists (amateur or professional), see Brock 2020.

10. These interactions repeat all day long, forming a sort of interpersonal representation of institutional neoliberalism and the intimate relationship it fosters between the carceral state and the welfare state.

11. *Patrons* is of course a generalization. While I spent most of my library fieldwork time with the homeless community, the Friends of the Library—wealthier, organized, politically astute—could and would make demands of the library's resources that were often granted: space for events, book sales, a conference. This of course did not happen as frequently as they might wish. At an October 2014 presentation by the National Capital Planning Commission addressing how the planned renovations to MLK would affect Mies van der Rohe's historic architecture, Caroline, the group's leader, muttered to me that new DCPL Executive Director Richard Reyes-Gavilan's shout out to the Friends was "window dressing" because the Friends were never consulted, just "talked at." Neither Caroline nor Bill—another Friend, passionate about teaching digital literacy—lived near MLK, nor did they regularly frequent its collections outside of meetings and events—a stark difference from other, admittedly better funded, Friends groups associated with specific branches.

12. The name is a recent coinage by the local Business Improvement District for the neighborhood north of Massachusetts Avenue NW, alien to DC natives.

4 Flexible Classrooms

1. Acumen Solutions is based in Northern Virginia and provides cloud-based analytics, storage, and customer relationship management services to firms in health, finance, media, and other fields. SchoolForce represented an important new push for them into enterprise-level services in public education. It stored, tracked, and analyzed standards-based gradebooks, homework data, behavioral data, attendance data, and even meals data.

2. These ideas emerge from the broad rubric of the reconceptualist movement in education. Reconceptualism shifts educators from the top-down implementation of universal curricula to the bottom-up implementation of curricula fitted to children's social context. Reconceptualism embraces feminist, antiracist, and democratic modes of education (Kessler and Swaddner 1992).

3. For example, Ryan attended Haverford College, whereas senior team leader Sam attended Harvard. Both entered the profession through Teach for America.

4. The senior class included two students who identified as Asian American and none who identified as White.

5. I did not witness the latter firsthand but had it confirmed to me in separate interviews. English teacher Melissa cited this, and more generally the stress that led teachers to this, as a major reason for her quitting mid-year.

6. Nationally, charters test about as well as traditional public schools (CREDO 2009; Gleason et al. 2010). Locally, the Rhee-era gains in test scores were largely discredited by what appeared to be widespread cheating by under-pressure teachers (Gillum and Bello 2011; Toppo 2013). In the first year of the new PARCC test, "7 percent of charter school students who took the high school Math test and 23 percent of those who took the high school English test scored proficient, compared with 12 and 27 percent of DC Public School students respectively" (Chandler 2015b).

7. In terms of disciplinary practices and racial climate, nationally, charters disproportionately suspend Black and disabled students and do so at a rate 16 percent higher than traditional public schools (Losen et al. 2016). To be fair, these patterns didn't hold in DC. There, the disparity between Black and disabled students and the median was maintained across the school system, but students in traditional public schools were 1.58 times more likely than charter school students to be suspended or expelled (OSSE 2013).

8. As evidence of this broken system, charter activists often cite DC's high per-pupil funding levels and middling standardized test scores relative to other states (see US Census Bureau 2019), though of course the comparison falls short because, unlike any state, DC is an entirely urban school district, the educational environment and student population of which looks more like, say, Chicago than Illinois (i.e., higher staff salaries and capital costs and poorer, more diverse students with higher needs).

9. Since then, DC has been superseded by cities such as New Orleans, which, in 2019, became an all-charter city with no traditional public schools (Hasselle 2018). The change is widely viewed as part of many neoliberal reforms instituted in the city following the shock of Hurricane Katrina and the subsequent redevelopment, including the elimination of public housing (Klein 2007).

10. Rhee was a Teach for America veteran tightly connected with that internship and advocacy organization. She founded the New Teacher Project (NTP) and, based on that work, was recommended to Fenty by New York City Schools Chancellor Joel Klein.

11. Netbooks are smaller and cheaper than traditional laptops, with less storage space, and are optimized for cloud-based software like Google Apps for Education.

12. Charters receive per-pupil funding from the state or county, but cannot raise taxes to support new capital expenditures like normal public school systems do through state and local governments. So they have pursued entrepreneurial fund-raising strategies, facilitated by looser governance structures and the heavy involvement of the financial services sector in lobbying for, funding, and, in individual cases, administering charters. Chief among these are (1) independent grant-writing efforts through dedicated fundraising offices and (2) bond offerings specific to the school and its special features, as distinct from the bond offerings conducted by whole public school systems. This financial creativity supports charter growth independent of student demand. Through 2014, nonprofits had donated $2.1 billion to charters, and charters gained $10 billion from the tax-exempt bond market. Charters' bond offerings assume constant growth for individual schools and are most often used to service existing debt (Teresa and Good 2018).

   In Du Bois's case, the school formed separate nonprofit ventures to secure financing for each school building (the middle school was separate from the high school in which I spent my time) and floated tax-exempt DC revenue bonds to support capital improvements and the refinancing of existing debt. Du Bois's per-student spending, one of the highest in the city and double a typical DCPS school, was boosted not just by these bond sales but also by grants and donations sought through a dedicated fundraising office. For example, in the summer of 2014, they were one of several schools awarded a $100,000 "transformational school" grant from the CityBridge Foundation to plan experiments in blended in-person and online learning, with up to $300,000 in additional funds available to support implementation. In DC, philanthropic donations are very unevenly distributed. In 2015, six charters, out of sixty total, received 75 percent of all donations to charter schools (Chandler 2015c).

13. The process of defining what academic success looks like and delimiting the good work habits that will get you there from the bad ones that will not has obvious racial and class overtones. For critics, this is a much remarked-on feature of the charter movement. A former dean of students at a New Orleans charter wrote: "My daily routine consisted of running around chasing young Black ladies to see if their nails were polished, or if they added a different color streak to their hair or following

young men to make sure that their hair wasn't styled naturally as students were not able to wear their hair in uncombed afro styles. None of which had anything to do with teaching and learning, but administration was keen on making sure that before Black students entered the classroom that they looked 'appropriate' for learning" (Griffin 2014; for a broader critique of respectability politics in education reform, see also Vasquez-Heilig, Khalifa, and Tillman 2014).

14. Study hall was at first an after-school catch-up program running at the same time as other extracurriculars like Robotics Club, open to anyone who could not or preferred not to work at home. Later, in another Du Bois experiment, it was required for struggling students. Attendance, however, was difficult to enforce.

5 Bootstrapping

1. CityBridge initially focused on global health before pivoting to education. Since the period covered here, the nonprofit split from its original founders to focus solely on education. The foundation focuses its school grants and teacher fellowships on DC public and charter schools, positioning itself as an entrepreneurial successor to local education reformers. They build on "the pioneering human capital reforms led by Michelle Rhee and Kaya Henderson in DCPS and from stand-out charter performers, like KIPP, DC Prep, and Two Rivers, whose work proves every day that the achievement gap can be closed." Like many venture philanthropists, their language and approach—"incubating, launching, and transforming schools and projects that will deliver on the talent and potential inherent in children, especially those disinherited by poverty or race"—is borrowed explicitly from the tech sector. See https://citybridge.org/about-us/.

NewSchools works nationally and explicitly operates as a venture capital fund modeled after its founders' experiences in the tech sector. They "were among the first and largest investors in public charter schools and the first to identify and support multi-site charter management organizations, which launch and operate integrated networks of public charter schools." (See http://www.newschools.org/about-us/our-history/.)

2. It is not uncommon for education reformers to compare the neoliberal process of education reform with Black freedom struggles. As Scott (2013) explains, these comparisons primarily serve to recast the story of the civil rights movement as one of individual struggles to secure individual rights to vote, shop, or choose a school. By removing these struggles from the context of Black political collectives, the neoliberal individualization of education policy is granted moral legitimacy and political urgency.

3. My use of the term *gospel* here is quite deliberate, a guiding principle that cannot be disproven and which inspires listeners to spread it further. In this usage, and throughout this book, I am indebted to Cottom's (2017) exploration of the *education*

gospel, a parallel phenomenon to the access doctrine that motivates the growth of the for-profit education sector (e.g., Phoenix, Strayer). Not just a sales pitch from for-profits or their financiers, the education gospel explains the poor labor market chances of working-class, particularly Black, American adults in terms of low educational attainment and justifies for-profit schools as a solution to this problem, not just despite but because of these schools' high costs—which, paradoxically, legitimates their degree offerings. This student population is often locked out of traditional avenues of higher education and, absent a robust welfare state and the sort of wages that would ensure a reasonable standard of living, welcome the advances of the heavily advertised for-profits that meet them where they are and help design degree programs that fit their working lives. Like the access doctrine, the education gospel explains how everyday, well-meaning people embrace a neoliberal project attempting to remake or destroy the very institutions they treasure as safe havens and guiding stars. Not because they are fooled, but because they must.

4. Many thanks to Victor Ray for helping me think through this point specifically and this literature more generally.

5. Teacher demographics across DC (including charter and traditional public schools, with dedicated data on the former difficult to acquire) roughly match what I observed in Du Bois. The overwhelming majority (75 percent) are women, and a slight majority (52 percent) are Black (OSSE 2019). If anything, there was a higher proportion of White teachers in Du Bois during my time there than there was in DC as a whole.

Demographic data for DC public librarians are harder to come by. At the national level, 2009–2010 American Community Survey data showed that 85 percent of public librarians are White, with that number dropping to 73.3 percent for library assistants (who do not have the MLIS). Seventy-five percent of the US population identified as White in the 2010 census (ALA 2010). If anything, I suspect DC's public librarians were more racially diverse, specifically with more Black librarians, than these national data indicate. But I have no systemic data to support this claim.

6. In contrast to its self-produced image as a haven for iconoclast dropouts such as Bill Gates or Mark Zuckerberg—dropouts from Harvard, to be clear—tech-sector job openings are more likely to list educational requirements than employers at large (Weber 2016). Chavez's (2017) case study of software engineer hiring at one Silicon Valley firm also demonstrates that hirers penalize foreign-born applicants with the necessary educational credentials and listed technological skills because of a perceived lack of "fit" with the culture of the US firm, a finding that would seem to contradict the data in Weber 2016. However, Silicon Valley's reliance on elite educational credentials *and* its willingness to ignore those credentials in favor of nativist arguments for fit both reveal a refusal to actually engage in the meritocracy the sector professes to embrace. When the chips are down, it's not what skills you have—it's which school you went to and what sort of parties you attend.

7. The imposition of this local meritocratic mindset onto larger social systems by White professionals is not dissimilar to how Ho (2009) describes investment bankers enforcing the local rules of financial institutions—long hours, frequent layoffs—in other spaces of economic life. In both cases, privilege is mistaken for hard work, not because hard work is absent but because its unceasing presence occludes larger structural factors. In both cases, there is a class project to enforce the rules of the game on other people who didn't even know they were playing.

8. One of the friendly fixtures of the Digital Commons was an older man who had previously taught at a local university but was now living rough. He spent much of his day reviewing books on Renaissance art and constitutional law, a good citizen of the library who still frightened some visitors who were not regulars with his constant, low-level self-talk. Many other homeless patrons had completed college or attempted some of it and dropped out. I have no comprehensive data on homeless patrons' educational attainment.

9. A 2017 survey of DC startups found that 15 percent were initially funded by founders' savings, 14 percent by friends and family, 19 percent through revenue, 9 percent through angel investors, and 9 percent through credit cards (Zuckerman et al. 2017). A mix of other, smaller funding sources (e.g., incubators, traditional loans) make up the remainder.

10. This conceptualization of institutional culture is heavily indebted to Thornton and Ocasio's (1999, 2008) approach to "institutional logics," particularly in their understanding of institutional logics as a metatheory that can explain social life at multiple levels of analysis, built on both material and cultural foundations, providing an overarching story without erasing individual agency or historical contingency. It also shares some similarities with Boltanski and Thevenot's (2006) approach to individual economic justifications of worth. I avoid using the term *institutional logics* for two reasons: First, because this chapter's model for organizational change is based on another branch of neoinstitutionalist theory (institutional isomorphism), which occurs at a lower level of analysis. Second, because I diverge from Thornton and Ocasio in my commitment to grounding that model for organizational change within a broader, structural theory of feminist Marxist political economy, which was discussed in the introduction and chapter 1 and which will be returned to in the conclusion. Institutional logics is thus both too big and too small a theoretical lens for this project's specific needs.

Conclusion

1. The crime rate certainly did begin to drop precipitously in the United States in the 1990s, but it did so throughout the world. So though the US carceral state played some role in lowering the crime rate by caging the working and workless poor—who, by virtue of the limited opportunities for the good life that result

from their labor market marginalization and the limited social services available in racially segregated areas, were more likely to both commit crimes and be arrested for them (Gilmore 2007)—this can hardly be the entire story. Testing seventeen different hypotheses from deleading gasoline to legalizing abortion, Farrell, Tilley, and Tseloni (2014) find that the only one that does not fail evidence-based standardized tests is the so-called security hypothesis: opportunity to commit property crimes in particular was reduced via better security systems in cars, homes, and businesses. The cause of the crime drop remains a matter of much debate, but what is clear is that the US federal government cannot take credit for it. Reductions in teenage pregnancies in the 1990s are easy to explain through increased access to family planning services, though in recent years the weak labor market (few jobs, low wages, or both) also seems to explain much of it (Kearney and Levine 2015). The reduction in welfare rolls is explained largely in the literature review on labor market boundary institutions in chapter 1: there was a multidecade effort to reduce the amount and availability of unemployment benefits, food stamps, and disability assistance that, in the 1990s, coincided with a strong labor market.

None of this is to disprove Clinton's, or my, point that the mode of social reproduction really was remade during the neoliberal revolution. Many new prisons were built and filled. Standardized testing and charter schools changed public education. Clinton really did "end welfare as we know it." But we cannot simply take elites at their word, especially when it is framed in such neutral terms, evacuated of political conflict. And there is no simple utilitarian calculus whereby new policy measures, especially those introduced by the federal government in a federalist system, generate a few new points, or not, of increase in prosocial outcomes. Indeed, one result of the neoliberal revolution is this understanding of policy not as a realm of collective struggle but as a series of dials turned up or down to produce measurable effects in discrete, unconnected individuals. The deeper effect, and, frequently, the real goal, of these reforms is to make new kinds of people with new attitudes, habits, and relationships to the institutions in their lives. I follow Wacquant (2009) in viewing many of these welfare state reforms as attempts to increase the labor market flexibility of the working and workless poor, disciplining them to accept low-wage service work and constant retraining and job search.

2. Although the empirical debate has greatly developed, the overall conceptual issues have not changed much in the past few decades, in part because defining skill is not a politically neutral affair—and, as Cappelli (2015) points out, employers are loathe to provide outside observers with detailed data on work tasks. For an overview of these conceptual issues, each article in *Work and Occupations* 17 (4), particularly Attewell 1990 and Steinberg 1990, is invaluable.

3. "Hegemony here is born in the factory and requires for its exercise only a minute quantity of professional and political intermediaries" (Gramsci 2000, 278–279). This does not mean that the political and cultural superstructure is mechanically determined by the economic base. For Marx, the base is not the assembly line but

the whole apparatus for reproducing capital and the workers it needs; and though the base limits the shape of and outcomes from the superstructure, it cannot hold its own shape without the support of particular legal, political, and cultural regimes (see Marx 1990, chapter 14, fn. 5, for the original theorization; see Williams 1973 for its most complete elaboration). For Gramsci, Fordism required Americanism, which meant "a particular social structure (or at least a determined intention to create it) and a certain type of state ... the liberal state, not in the sense of free-trade liberalism or of effective political liberty, but in the more fundamental sense of free initiative and of economic individualism" (285). The baseline assumption here is that people are intelligent, creative political actors and their cooperation in a project that might not be in some of their interests must be secured via an appeal to some other of them.

4. This is not to say that these movements were wholly professional ones. They were not, despite the mythology around revanchist working-class protests like the 1970 Hard Hat Riot, wherein construction workers attacked Vietnam War protestors. See Windham 2017 for a critical reassessment of the relationship between the working class of the late 1960s and 1970s and these movements.

References

Adamson, Morgan. 2009. "The Human Capital Strategy." *Ephemera* 9 (4): 271–284.

AFSCME (American Federation of State, County, and Municipal Employees). 1968. "Librarian Members of AFSCME Local 1675, Contra Costa & Solano County, California, Picket in August, 1968." Walter P. Reuther Library, Wayne State University. http://reuther.wayne.edu/node/2079.

Agosto, D. E. 2008. "Alternative Funding for Public Libraries: Trends, Sources, and the Heated Arguments That Surround It." In *Influence of Funding on Advances in Librarianship*, edited by Danuta A. Nitecki and Eileen E. Abels, 115–140. Bingley, UK: Emerald Group Publishing Limited.

Akee, Randall, Maggie R. Jones, and Sonya R. Porter. 2019. "Race Matters: Income Shares, Income Inequality, and Income Mobility for All US Races." *Demography* 56 (3): 999–1021.

Allard, Scott W. 2007. "The Changing Face of Welfare during the Bush Administration." *Publius: The Journal of Federalism* 37 (3): 304–332.

Anderson, Ben, and Adam Holden. 2008. "Affective Urbanism and the Event of Hope." *Space and Culture* 11 (2): 142–159.

Anyon, Jean. 1980. "Social Class and the Hidden Curriculum of Work." *Journal of Education* 162 (1): 67–92.

Archibald, Robert B., and David H. Feldman. 2006. "State Higher Education Spending and the Tax Revolt." *Journal of Higher Education* 77 (4): 618–644. https://doi.org/10.1080/00221546.2006.11772309.

Asaro, Peter M. 2000. "Transforming Society by Transforming Technology: The Science and Politics of Participatory Design." *Accounting, Management and Information Technologies* 10 (4): 257–290.

Attewell, Paul. 1990. "What Is Skill?" *Work and Occupations* 17 (4): 422–448.

Audretsch, Bruce. 1998. "Agglomeration and the Location of Innovative Activity." *Oxford Review of Economic Policy* 14 (2): 18–29.

Autor, David H., and David Dorn. 2013. "The Growth of Low-Skill Service Jobs and the Polarization of the US Labor Market." *American Economic Review* 103 (5): 1553–1597.

Autor, David H., Lawrence F. Katz, and Alan B. Krueger. 1998. "Computing Inequality: Have Computers Changed the Labor Market?" *Quarterly Journal of Economics* 113 (4): 1169–1213.

Baily, Martin Neil, and Nicholas Montalbano. 2016. "Why Is US Productivity Growth So Slow? Possible Explanations and Policy Responses." Brookings Institution, Hutchins Center Working Paper 22. http://www.nomurafoundation.or.jp/word press/wp-content/uploads/2016/12/20161116_M_Baily-N_Montalbano.pdf.

Bakker, Isabella. 2007. "Social Reproduction and the Constitution of a Gendered Political Economy." *New Political Economy* 12 (4): 541–556.

Beck, Eevi E. 2002. "P for Political: Participation Is Not Enough." *Scandinavian Journal of Information Systems* 14 (1): 77–92.

Benanav, Aaron. 2014. "A Global History of Unemployment." PhD diss., University of California, Los Angeles.

Berger, Dan. 2013. "Social Movements and Mass Incarceration: What Is to Be Done?" *Souls* 15 (1–2): 3–18.

Berghel, Hal. 2014. "STEM, Revisited." *Computer* 47 (3): 70–73.

Bertot, John Carlo, Abigail McDermott, Ruth Lincoln, Brian Real, and Kaitlin Peterson. 2012. *2011–2012 Public Library Funding and Technology Access Survey: Survey Findings and Results*. College Park, MD: Information Policy & Access Center, University of Maryland College Park. http://www.ala.org/tools/research/plftas/2011_2012.

Besharov, Marya L., and Wendy K. Smith. 2014. "Multiple Institutional Logics in Organizations: Explaining Their Varied Nature and Implications." *Academy of Management Review* 39 (3): 364–381.

Bezanson, Kate. 2006. *Gender, the State, and Social Reproduction: Household Insecurity in Neo-liberal Times*. Toronto: University of Toronto Press.

Bhattacharya, Tithi. 2020. "Liberating Women from 'Political Economy': Margaret Bentson's Marxism and a Social-Reproduction Approach to Gender Oppression." *Monthly Review*, January 1, 2020. https://monthlyreview.org/2020/01/01/liberating -women-from-political-economy/?v=7516fd43adaa.

Biddle, RiShawn. 2009. "The BlackBerry Kid: Michelle Rhee's Showdown with the DC Teachers Union." *CRC Labor Watch*, January 2009. https://capitalresearch.org /app/uploads/2013/07/LW0109.pdf.

Bivens, Josh, Elise Gould, Lawrence R. Mishel, and Heidi Shierholz. 2014. "Raising America's Pay: Why It's Our Central Economic Policy Challenge." Economic Policy Institute. https://www.epi.org/publication/raising-americas-pay/.

Blanchard, Dana. 2018. "Lessons from the Teachers' Strike Wave." *International Socialist Review*, no. 110. https://isreview.org/issue/110/lessons-teachers-strike-wave.

Bloom, Howard S., Larry L. Orr, Stephen H. Bell, George Cave, Fred Doolittle, Winston Lin, and Johannes M. Bos. 1997. "The Benefits and Costs of JTPA Title II-A Programs: Key Findings from the National Job Training Partnership Act Study." *Journal of Human Resources* 32 (3): 549–576.

Bloomberg News. 2007. "Greenspan: Let More Skilled Immigrants In." *Boston Globe*, March 14. http://archive.boston.com/business/globe/articles/2007/03/14/greenspan_let_more_skilled_immigrants_in/.

BLS (US Bureau of Labor Statistics). 2015a. "STEM Crisis or STEM Surplus? Yes and Yes." Monthly Labor Review, May 2015. http://www.bls.gov/opub/mlr/2015/article/stem-crisis-or-stem-surplus-yes-and-yes.htm.

BLS (US Bureau of Labor Statistics). 2015b. "Occupations with the Most Job Growth." Employment Projections, December 8, 2015. https://www.bls.gov/emp/tables/occupations-most-job-growth.htm.

BLS (US Bureau of Labor Statistics). 2016. *Employment Experience of Youths: Results from a Longitudinal Survey News Release.* Washington, DC: United States Department of Labor, April 8. http://www.bls.gov/news.release/nlsyth.htm.

Bonhomme, Edna. 2016. "Behind the Lockout: An Interview with Emily Drabinski." *Jacobin*, September 15, 2016. https://www.jacobinmag.com/2016/09/behind-the-lockout/.

Boltanski, Luc, and Laurent Thévenot. 2006. *On Justification: Economies of Worth.* Princeton: Princeton University Press.

Boothe, Charlie. 2018. Potential Teacher Strike Looms over West Virginia. *Bluefield Daily Telegraph*, January 29, 2018. https://www.bdtonline.com/news/potential-teacher-strike-looms-over-west-virginia/article_32f4a9f4-04a1-11e8-99f2-7f31dc816267.html.

Bowser, Muriel, and Jeffrey S. DeWitt. 2018. *Review of Economic Development Tax Expenditures.* Office of Revenue Analysis. Washington, DC: Government of the District of Columbia. https://cfo.dc.gov/sites/default/files/dc/sites/ocfo/publication/attachments/2017%20Economic%20Development%20Tax%20Expenditures%20121918.pdf.

Bowser, Muriel, Jeffrey S. De Witt, and Fitzroy Lee. 2015. "DC's Population Grew for the 9th Straight Year in 2014, but Growth Was the Slowest in 6 Years." DC Office of Revenue Analysis Briefing Document 2015-1. http://cfo.dc.gov/sites/default/files/dc/sites/ocfo/publication/attachments/2015-01_DC%20Population_0.pdf.

Brock, André, Jr. 2020. *Distributed Blackness: African American Cybercultures.* New York: NYU Press.

Brodkin, Jon. 2018. "FCC Republican Claims Municipal Broadband Is Threat to First Amendment." *Ars Technica* (blog), October 30, 2018. https://arstechnica.com /tech-policy/2018/10/fcc-republican-claims-municipal-broadband-is-threat-to-first -amendment/.

Brown, L. David. 1991. "Bridging Organizations and Sustainable Development." *Human Relations* 44 (8): 807–831.

Bush, George W. 2004. "President Unveils Tech Initiatives for Energy, Health Care, Internet: Remarks by the President at American Association of Community Colleges Convention." Office of the Press Secretary, April 26, 2004. https://georgewbush -whitehouse.archives.gov/news/releases/2004/04/text/20040426-6.html.

Calarco, Jessica McCrory. 2018. *Negotiating Opportunities: How the Middle Class Secures Advantages in School.* New York: Oxford University Press.

Callon, Michel. 1984. "Some Elements of a Sociology of Translation: Domestication of the Scallops and the Fishermen of St. Brieuc Bay." *Sociological Review* 32 (1): 196–233.

Camp, Jordan T. 2016. *Incarcerating the Crisis: Freedom Struggles and the Rise of the Neoliberal State.* Oakland: University of California Press.

Cappelli, Peter H. 1999. "Career Jobs Are Dead." *California Management Review* 42 (1): 146–167.

Cappelli, Peter H. 2015. "Skill Gaps, Skill Shortages, and Skill Mismatches: Evidence and Arguments for the United States." *ILR Review* 68 (2): 251–290.

Cate, Fred H. 1994. "The National Information Infrastructure: Policymaking and Policymakers." *Stanford Law and Policy Review* 6 (1): 43–60.

Carragee, Kevin M., and Wim Roefs. 2004. "The Neglect of Power in Recent Framing Research." *Journal of Communication* 54 (2): 214–233.

Castaneda, Ruben. 2020. "What's The Plan?" *Washington City Paper*, March 5, 2020. https://www.washingtoncitypaper.com/news/article/21120072/whats-the-plan.

Chandler, Michael Alison. 2015a. "DC Charter Schools Serve Fewer At-Risk Students Than Nearby Neighborhood Schools." *Washington Post*, October 8, 2015. https://www .washingtonpost.com/news/education/wp/2015/10/08/d-c-charter-schools-serve -fewer-at-risk-students-than-nearby-neighborhood-schools/.

Chandler, Michael Alison. 2015b. "College Readiness Scores Range Widely at DC High Schools." *Washington Post*, October 28, 2015. https://www.washingtonpost .com/news/education/wp/2015/10/28/college-readiness-scores-range-widely-at-d-c -high-schools/.

Chandler, Michael Alison. 2015c. "Some DC Charter Schools Get Millions in Dona- tions; Others, Almost Nothing." *Washington Post*, August 26, 2015. https://www.wash ingtonpost.com/local/education/some-charter-schools-get-millions-in-donations

-others-almost-nothing/2015/08/22/b1fdaef0-4804-11e5-8e7d-9c033e6745d8_story
.html?hpid=z4.

Chaplan, Margaret A. 1976. "Appendix A: Chronology of Job Actions, 1966–1975."
In Employee Organization and Collective Bargaining in Libraries, edited by Margaret
A. Chaplan. *Library Trends* 25 (2): 517–524.

Charette, Robert N. 2013. "The STEM Crisis Is a Myth." *IEEE Spectrum* 50 (9): 44–59.

Chavous, Kevin. 2009. "Charter Schools. ICn *Mandate for Change*, edited by Samuel
C. Carter, 19–25. Washington, DC: Center for Education Reform.

Clinton, William J. 1993. "Address before a Joint Session of Congress on Administration Goals." American Presidency Project, February 17, 1993. https://www.presidency
.ucsb.edu/documents/address-before-joint-session-congress-administration-goals.

Clinton, William J. 2000a. "Address before a Joint Session of the Congress on the State
of the Union." American Presidency Project, January 27, 2000. https://www.presidency
.ucsb.edu/documents/address-before-joint-session-the-congress-the-state-the-union-7.

Clinton, William J. 2000b. "Remarks by the President on Digital Opportunities for
Americans with Disabilities." Office of the Press Secretary, September 21, 2000. http://
www.icdri.org/DD/clintonddspeech.htm.

Clinton, William J., and the Council of Economic Advisors. 1994. *Economic Report of
the President Transmitted to the Congress, Together with the Annual Report of the Council
of Economic Advisers*. Washington, DC: United States Government Printing Office.

Clinton, William J., and Al Gore Jr. 1993. "Technology for America's Economic
Growth: A New Direction for Building Economic Strength." Office of the Press Secretary, February 22, 1993.

Clinton, William J., and Al Gore Jr. 1995. "A Framework for Global Electronic Commerce." Office of the Press Secretary, July 1, 1997.

Clinton, William J., and Al Gore Jr. 1996a. "Background on the Clinton-Gore
Administration's Next-Generation Internet Initiative." Office of the Press Secretary,
October 10, 1996.

Clinton, William J., and Al Gore Jr. 1996b. "Remarks by the President and Vice President to the People of Knoxville." Office of the Press Secretary, October 10, 1996.

Cobble, D. S. 1991. "Organizing the Postindustrial Work Force: Lessons from the
History of Waitress Unionism." *ILR Review* 44 (3): 419–436.

Cottom, Tressie McMillan. 2016. "Black Cyberfeminism: Ways Forward for Intersec-
tionality and Digital Sociology." In *Digital Sociologies*, edited by Jessie Daniels, Karen
Gregory, and Tressie McMillan Cottom, 211–232. Bristol, UK: Polity Press.

Cottom, Tressie McMillan. 2017. *Lower Ed: The Troubling Rise of For-Profit Colleges in
the New Economy*. New York: New Press.

Craig, Tim, and Bill Turque. 2010. "Michelle Rhee Resigns; Gray Huddles with Her Successor." *Washington Post*, October 13, 2010. http://www.washingtonpost.com/wp -dyn/content/article/2010/10/12/AR2010101205658.html.

Crawford, Susan. 2013. *Captive Audience: The Telecom Industry and Monopoly Power in the New Gilded Age*. New Haven: Yale University Press.

CREDO (Center for Research on Education Outcomes). 2009. *Multiple Choice: Charter School Performance in 16 States*. Stanford University, Stanford, CA. https://credo .stanford.edu/sites/g/files/sbiybj6481/f/multiple_choice_credo.pdf.

Dalla Costa, Mariarosa, and Selma James. 1975. *The Power of Women and the Subversion of the Community*. Bristol: Falling Wall Press.

Dani, Lokesh. 2013. "The Lower-Wage Recovery in the Higher-Wage Economy of the Washington, DC Metropolitan Area." George Mason University Center for Regional Analysis, Working Paper No. 2013-07.

Darity, William. 1983. "The Managerial Class and Surplus Opulation." *Society* 21 (1): 54–62.

DCPCSB (DC Public Charter School Board). 2020. "Historical Enrollment Analysis." OpenData DC PCSB, July 15, 2020. https://dcpcsb.egnyte.com/dl/0BR9yldipC/.

Dean, Jodi. 2008. "Enjoying Neoliberalism." *Cultural Politics* 4 (1): 47–72.

Department of Commerce. 1993. *The National Information Infrastructure: Agenda for Action*. Washington, DC: Information Infrastructure Task Force. http://www.ibiblio .org/nii/toc.html.

DeVault, Marjorie L. 2006. "Introduction: What Is Institutional Ethnography?" *Social Problems* 53 (3): 294–298.

DiMaggio, Paul. 1991. "Constructing an Organizational Field as a Professional Project: The Case of U.S. Art Museums." In *The New Institutionalism in Organizational Analysis*, edited by Walter W. Powell and Paul J. DiMaggio, 267–292. Chicago: University of Chicago Press.

DiMaggio, Paul, Ezster Hargittai, Coral Celeste, and Steven Shafer. 2004. "Digital Inequality: From Unequal Access to Differentiated Use." In *Social Inequality*, edited by Katheryn Neckerman, 355–400. New York: Russell Sage Foundation.

DiMaggio, Paul, and Walter W. Powell. 1983. "The Iron Cage Revisited: Institutional Isomorphism and Collective Rationality in Organizational Fields." *American Sociological Review* 48 (2): 147–160.

DMPED (Deputy Mayor for Planning & Economic Development). 2016. "Economic Intelligence Dashboard: Employment and Labor, March 2016." http://open.dc.gov /economic-intelligence/employment-labor.html.

Drucker, Peter F. 1999. "Knowledge-Worker Productivity: The Biggest Challenge." *California Management Review* 41 (2): 79–94.

Duménil, Gérard, and Dominique Lévy. 2015. "Neoliberal Managerial Capitalism: Another Reading of the Piketty, Saez, and Zucman Data." *International Journal of Political Economy* 44 (2): 71–89.

Duster, Troy. 1995. "The New Crisis of Legitimacy in Controls, Prisons, and Legal Structures." *American Sociologist* 26 (1): 20–29.

Edelman, Peter. 2013. *So Rich, So Poor: Why It's So Hard to End Poverty in America*. New York: New Press.

Ehrenreich, Barbara, and John Ehrenreich. 1977. "The Professional-Managerial Class." *Radical America*, March-April, 7–31.

Ehrenreich, Barbara, and John Ehrenreich. 2013. *Death of a Yuppie Dream: The Rise and Fall of the Professional-Managerial Class*. New York: Rosa Luxemburg Siftung. http://www .rosalux-nyc.org/wp-content/files_mf/ehrenreich_death_of_a_yuppie_dream90.pdf.

Enjeti, Saagar. 2019. "Ride Is Likely over for Kamala Harris." *The Hill*, October 9, 2019. https://thehill.com/hilltv/rising/465008-saagar-enjeti-ride-is-likely-over-for-kamala -harris.

Epstein, Dmitry, Erick Nisbet, and Tarleton Gillespie. 2011. "Who's Responsible for the Digital Divide? Public Perceptions and Policy Implications." *Information Society*, 27 (2): 92–104.

Escobar, Arturo. 2011. *Encountering Development: The Making and Unmaking of the Third World*. Princeton: Princeton University Press.

Eubanks, Virginia. 2006. "Technologies of Citizenship: Surveillance and Political Learning in the Welfare System." In *Surveillance and Security: Technology and Power in Everyday Life*, edited by Torin Monahan, 89–108. New York: Routledge.

Eubanks, Virginia. 2007. "Trapped in the Digital Divide: The Distributive Paradigm in Community Informatics." *Journal of Community Informatics* 3 (2). http://ci-journal .net/index.php/ciej/article/view/293.

Eubanks, Virginia. 2011. *Digital Dead End: Fighting for Social Justice in the Information Age*. Cambridge, MA: MIT Press.

Evans, William N., Robert M. Schwab, and Kathryn L. Wagner. 2019. "The Great Recession and Public Education." *Education Finance and Policy* 14 (2): 298–326.

Farrell, Graham, Nick Tilley, and Andromachi Tseloni. 2014. "Why the Crime Drop?" *Crime and Justice* 43 (1): 421–490.

FCC (Federal Communications Commission). 1997. "Report & Order in the Matter of Federal-State Joint Board on Universal Service." CC docket no. 96-45.

FCC (Federal Communications Commission). 2016. *Broadband Progress Report*. Washington, DC: Federal Communications Commission.

Federici, Silvia. 2004 *Caliban and the Witch: Women, the Body, and Primitive Accumulation*. Brooklyn: Autonomedia.

Ferguson, James. 1994. *The Anti-politics Machine: "Development," Depoliticization, and Bureaucratic Power in Lesotho*. Minneapolis: University of Minnesota Press.

Ferguson, Thomas. 1995. *Golden Rule: The Investment Theory of Party Competition and the Logic of Money-Driven Political Systems*. Chicago: University of Chicago Press.

Fernandez, Manny. 2004. "Seeking to Turn a Page on Disrepair: After Languishing for Years, the District's Libraries Are Getting New Attention." *Washington Post*, April 8, 2004, sec. DE 10.

Florida, Richard. 2004. *The Rise of the Creative Class*. New York: Basic Books.

Forman Jr., James. 2017. *Locking Up Our Own: Crime and Punishment in Black America*. New York: Farrar, Straus and Giroux.

Fortunati, Leopoldina. 1995. *The Arcane of Reproduction: Housework Prostitution, Labor and Capital*. Brooklyn: Autonomedia.

Fox, Melodie, and Hope Olson. 2013. "Essentialism and Care in a Female-Intensive Profession." In *Feminist and Queer Information Studies Reader*, edited by Patrick Keilty and Rebecca Dean, 48–61. Sacramento: Litwin Books.

Fraser, Nancy. 1993. "Clintonism, Welfare, and the Antisocial Wage: The Emergence of a Neoliberal Political Imaginary." *Rethinking Marxism* 6 (1): 9–23.

Fraser, Nancy. 2016. "Contradictions of Capital and Care." *New Left Review* 100 (99): 117.

Fraser, Nancy, and Linda Gordon. 1994. "A Genealogy of Dependency: Tracing a Keyword of the US Welfare State." *Signs* 19 (2): 309–336.

Friedland, Roger, and Robert R. Alford. 1991. "Bringing Society Back in: Symbols, Practices, and Institutional Contradictions." In *The New Institutionalism in Organizational Analysis*, edited by Walter W. Powell and Paul J. DiMaggio, 232–263. Chicago: University of Chicago Press.

Gates, Bill. 2005. "What's Wrong with American High Schools." *Los Angeles Times*, March 1, 2005. https://www.latimes.com/archives/la-xpm-2005-mar-01-oe-gates1-story.html.

Garrison, Dee. 1979. *Apostles of Culture: The Public Librarian and American Society, 1876–1920*. Madison: University of Wisconsin Press.

Geismer, Lily. 2014. *Don't Blame Us: Suburban Liberals and the Transformation of the Democratic Party*. Princeton: Princeton University Press.

Geismer, Lily. 2019. "Let Them Eat Tech." *Dissent* 66 (4): 48–57.

Giachetti, Steven. 2015. "Is CEO Pay the Major Cause of Income Inequality in the District? Increasingly the Corporate Ladder You're on Matters More than Where You Are on the Ladder." *District, Measured* (blog), District of Columbia Office of Revenue Analysis, June 23, 2015. http://districtmeasured.com/2015/06/23/ceo-pay-is-not-the-only-factor-that-accounts-for-income-inequality-in-dc-increasingly-the-corporate-ladder-youre-on-matters-more-than-where-you-are-on-the-corporate-ladder/.

Gillum, Jack, and Marisol Bello. 2011. "When Standardized Test Scores Soared in DC, Were the Gains Real?" *USA Today*, March 30, 2011. http://usatoday30.usatoday.com/news/education/2011-03-28-1Aschooltesting28_CV_N.htm.

Gilmore, Ruth Wilson. 1999. "Globalization and US Prison Growth: From Military Keynesianism to Post-Keynesian Militarism." *Race & Class* 40 (2/3): 171–188.

Gilmore, Ruth Wilson. 2007. *Golden Gulag: Prisons, Surplus, Crisis, and Opposition in Globalizing California*. Oakland: University of California Press.

Gilmore, Ruth Wilson, and Craig Gilmore. 2008. "Restating the Obvious." In *Indefensible Space: The Architecture of the National Insecurity State*, edited by Michael Sorkin, 141–62. New York: Routledge.

Giroux, Henry, and Anthony N. Penna. 1979. "Social Education in the Classroom: The Dynamics of the Hidden Curriculum." *Theory & Research in Social Education* 7 (1): 21–42.

Glaeser, Andreas. 2005. "An Ontology for the Ethnographic Analysis of Social Processes: Extending the Extended-Case Method." *Social Analysis* 49 (3): 16–45.

Gleason, Philip, Melissa Clark, Christina Clark Tutle, Emily Dwoyer, and Marshal Silverberg. 2010. *The Evaluation of Charter School Impacts: Final Report (NCEE 2010-4029)*. Washington, DC: National Center for Education Evaluation and Regional Assistance, Institute of Education Sciences, US Department of Education.

Golan, Joanne W. 2015. "The Paradox of Success at a No-Excuses School." *Sociology of Education* 88 (2): 103–119.

Goldhader, Dan D., Dominic J. Brewer, and Deborah J. Anderson. 1999. "A Three-Way Error Components Analysis of Educational Productivity." *Education Economics* 7 (3): 199–208.

Goldin, Claudia, and Lawrence Katz. 2008. *The Race between Technology and Education*. Cambridge, MA: Harvard University Press.

Gonzales, Amy. 2016. "The Contemporary US Digital Divide: From Initial Access to Technology Maintenance." *Information, Communication & Society* 19 (2): 234–248.

Gordon, Robert J. 2017. *The Rise and Fall of American Growth: The US Standard of Living since the Civil War*. Princeton: Princeton University Press.

Gore, Al, Jr. 1991. "Infrastructure for the Global Village." *Scientific American* 265 (3): 150–153.

Gore, Al, Jr. 1994a. "No More Information Haves and Have-Nots." *Billboard* 106 (43): 6.

Gore, Al, Jr. 1994b. "Remarks (as Delivered) by Vice President Al Gore via Satellite to the International Telecommunication Union Plenipotentiary Conference." Office of the Vice President, September 22, 1994. https://clintonwhitehouse6.archives.gov /1994/09/1994-09-22-vp-al-gore-via-satellite-to-itu-conference-kyoto.html.

Gore, Al, Jr. 1994c. "Remarks Delivered at UCLA Television Academy." January 11. http://www.ibiblio.org/icky/speech2.html.

Gore, Al, Jr. 1997. "Remarks by Vice President Al Gore at the Microsoft CEO Summit." Office of the Vice President, May 8. https://clintonwhitehouse2.archives.gov/WH/EOP /OVP/speeches/microsof.html.

Gowan, Teresa. 2009. "New Hobos or Neo-romantic Fantasy? Urban Ethnography beyond the Neoliberal Disconnect." *Qualitative Sociology* 32 (3): 231–257.

Gowan, Teresa. 2010. *Hobos, Hustlers, and Backsliders: Homeless in San Francisco.* Minneapolis: University of Minnesota Press.

Graham, Mark. 2008. "Warped Geographies of Development: The Internet and Theories of Economic Development." *Geography Compass* 2 (3): 771–789.

Gramsci, Antonio. 2000. *The Antonio Gramsci Reader: Selected Writings 1916–1935.* Edited by David Forgacs. New York: New York University Press.

Gray, M. L., and S. Suri. 2019. *Ghost Work: How to Stop Silicon Valley from Building a New Global Underclass.* New York: Houghton Mifflin Harcourt.

Green, Elizabeth. 2016. "Beyond the Viral Video: Inside Educators' Emotional Debate about 'No Excuses' Discipline." *Chalkbeat New York* (blog), March 8, 2016. http://ny.chalkbeat.org/2016/03/08/beyond-the-viral-video-inside-educators -emotional-debate-about-no-excuses-discipline/#.Vuss0xg7Tyn.

Gregg, Melissa. 2013. *Work's Intimacy.* New York: John Wiley & Sons.

Griffin, Ramon. 2014. "Colonizing the Black Natives: Reflections from a Former NOLA Charter School Dean of Students." *Cloaking Inequity* (blog), March 24, 2014. http://cloakinginequity.com/2014/03/24/colonizing-the-black-natives-reflections -from-a-former-nola-charter-school-dean-of-students/.

Gruber, Frank. 2014. *Startup Mixology: Tech Cocktail's Guide to Building, Growing, and Celebrating Startup Success.* New York: John Wiley & Sons.

Gunkel, David J. 2003. "Second Thoughts: Toward a Critique of the Digital Divide." *New Media & Society* 5 (4): 499–522.

Gunn, Richard. 1987. "Notes on 'Class.'" *Common Sense* 2:15–26.

Hacker, Jacob S. 2019. *The Great Risk Shift: The New Economic Insecurity and the Decline of the American Dream*. New York: Oxford University Press.

Hall, Stuart, Chas Critcher, Tony Jefferson, John Clarke, and Brian Roberts. 2013. *Policing the Crisis: Mugging, the State and Law and Order*. New York: Palgrave Macmillan.

Hamilton, Darrick, William Darity Jr., Anne E. Price, Vishnu Sridharan, and Rebecca Tippett. 2015. *Umbrellas Don't Make It Rain: Why Studying and Working Hard Isn't Enough for Black Americans*. New York: New School.

Hammond, Allen S., IV. 1997. "The Telecommunications Act of 1996: Codifying the Digital Divide." *Federal Communications Law Journal* 50 (1): 179–214.

Hancock, Ange-Marie. 2004. *The Politics of Disgust: The Public Identity of the Welfare Queen*. New York: NYU Press.

Hannerz, Ulf. 2003. "Being There … and There … and There! Reflections on Multi-site Ethnography." *Ethnography* 4 (2): 201–216.

Haraway, Donna J. 1997. *Modest_Witness@ Second_Millennium. FemaleMan©_Meets_ OncoMouse™*. New York: Routledge.

Hardt, Michael. 2005. "Immaterial Labor and Artistic Production." *Rethinking Marxism* 17 (2): 175–177.

Hargittai, Eszter, and Amanda Hinnant. 2008. "Digital Inequality Differences in Young Adults' Use of the Internet." *Communication Research* 35 (5): 602–621.

Harris, Donna. 2016. "Hope for the Future: Why I'm Stepping Down from 1776." *Startup Grind* (blog), Medium, November 23, 2016. https://medium.com/startup -grind/hope-for-the-future-why-im-stepping-down-from-1776-709f95d2f0cb.

Harvey, Charles, Mairi Maclean, Jillian Gordon, and Eleanor Shaw. 2011. "Andrew Carnegie and the Foundations of Contemporary Entrepreneurial Philanthropy." *Business History* 53 (3): 425–450.

Harvey, David. 1989. "From Managerialism to Entrepreneurialism: The Transformation in Urban Governance in Late Capitalism." *Geografiska Annaler: Series B, Human Geography* 71 (1): 3–17.

Hasselle, Della. 2018. "After McDonough 35 Vote, New Orleans Will Be 1st in US without Traditionally Run Public Schools." *New Orleans Times-Picayune*, December 20, 2018.

Henwood, Doug. 2019. "There Are So Many Things that We Can Learn from This Strike: An Interview with Alex Caputo-Pearl and Jane McAlevey." *Jacobin*, February 8, 2019.

Hochschild, Arlie Russell. 2012 *The Managed Heart: Commercialization of Human Feeling*. Second edition. Berkeley: University of California Press.

HUD (US Department of Housing and Urban Development). 2014. "2007–2014 Point in Time Counts by Continuum of Care." Homelessness Data Exchange. https://www.hudexchange.info/resources/documents/2007-2014-PIT-Counts-by-CoC.xlsx.

Hyra, Derek, and Sabihya Prince, eds. 2016. *Capital Dilemma: Growth and Inequality in Washington, DC*. New York: Routledge.

Ifill, Gwen. 1992. "Clinton's Standard Speech: A Call for Responsibility." *New York Times*, April 26, 1992.

IMLS (Institute of Museums and Library Services). 2017. *State Library and Administrative Agencies Survey: Fiscal Year 2016*. Washington, DC: Institute of Museum and Library Services.

Irani, Lilly. 2019. *Chasing Innovation: Making Entrepreneurial Citizens in Modern India*. Princeton: Princeton University Press.

Isaac, Mike, and Katie Benner. 2015. "LivingSocial Offers a Cautionary Tale to Today's Unicorns." *New York Times*, November 20, 2015. https://www.nytimes.com/2015/11/22/technology/livingsocial-once-a-unicorn-is-losing-its-magic.html?_r=0&referer=http://techmeme.com/m/.

Jackson, Esther Cooper. 1940. "The Negro Woman Domestic Worker in Relation to Trade Unionism." Master's thesis, Oberlin College. https://www.viewpointmag.com/2015/10/31/the-negro-woman-domestic-worker-in-relation-to-trade-unionism-1940/.

James, Nicky. 1989. "Emotional Labour: Skill and Work in the Social Regulation of Feelings." *Sociological Review* 37 (1): 15–42.

Jansa, Joshua M. 2020. "Chasing Disparity: Economic Development Incentives and Income Inequality in US States." *State Politics & Policy Quarterly*. https://doi.org/10.1177/1532440019900259.

Jayakar, Krishna, and Eun-A Park. 2012. "Funding Public Computing Centers: Balancing Broadband Availability and Expected Demand." *Government Information Quarterly* 29 (1): 50–59.

Katz, Cindi. 1998. "Excavating the Hidden City of Social Reproduction: A Commentary." *City & Society* 10 (1): 37–46.

Katz, Cindi. 2001. "Vagabond Capitalism and the Necessity of Social Reproduction." *Antipode* 33 (4): 709–728.

Katznelson, Ira. 2013. *Fear Itself: The New Deal and the Origins of Our Time*. New York: W. W. Norton & Company.

Kennicott, Philip. 2020. "America's Libraries Are Essential Now—and This Beautifully Renovated One in Washington Gives Us Hope." *Washington Post*, July 15, 2020. https://www.washingtonpost.com/entertainment/americas-libraries-are-essential

-now--and-this-beautifully-renovated-one-in-washington-gives-us-hope/2020/07/15 /d7b0bbc6-c5ed-11ea-b037-f9711f89ee46_story.html.

Kernochan, Rose. 2016. "Fundraising in American Public Libraries: An Overview." *Serials Librarian* 71 (2): 132–137.

Kessler, Shirley A., and Beth B. Swadener, eds. 1992. *Reconceptualizing the Early Childhood Curriculum: Beginning the Dialogue.* New York: Teachers College Press.

Khazan, Olga. 2015. "The Sexism of Startup Land." *Atlantic*, March 12, 2015. http:// www.theatlantic.com/business/archive/2015/03/the-sexism-of-startup-land/387184/.

Khimm, Suzy. 2014. "In DC, Inequality Hits Home." MSNBC, January 27, 2014. http://www.msnbc.com/msnbc/dc-inequality-hits-home.

Kim, Jae-Young. 1998. "Universal Service and Internet Commercialization: Chasing Two Rabbits at the Same Time." *Telecommunications Policy* 22 (4–5): 281–288.

Kinney, Bo. 2010. "The Internet, Public Libraries, and the Digital Divide." *Public Library Quarterly* 29 (2): 104–161.

Klein, Naomi. 2007. *The Shock Doctrine: The Rise of Disaster Capitalism.* New York: Macmillan.

Koebler, J. 2015. "The 21 Laws States Use to Crush Broadband Competition." *Motherboard*, January 11, 2015. https://motherboard.vice.com/en_us/article/qkvn4x/the-21 -laws-states-use-to-crush-broadband-competition.

Kolodny, Lora. 2014. "Education Investors, Startups Hopeful Following Solid IPO by 2U." *Wall Street Journal*, April 3, 2014. http://blogs.wsj.com/venturecapital/2014/04 /03/education-investors-startups-hopeful-following-solid-ipo-by-2u/.

Krueger, Alan B. 1993. "How Computers Have Changed the Wage Structure: Evidence From Microdata, 1984–1989." *Quarterly Journal of Economics* 108(1): 33–60.

Kuhnhenn, Jim. 2015. "Obama Calls for Effort to Boost High-Tech Training, Hiring." Associated Press, March 10, 2015. https://apnews.com/f124001ef99248a2b971419bd9 9736f1.

Labaton, Stephen. 2001. "New FCC Chief Would Curb Agency Reach." *New York Times*, February 7, 2001. http://www.nytimes.com/2001/02/07/business/new-fcc-chief -would-curb-agency-reach.html.

Lack, Brian. 2009. "No Excuses: A Critique of the Knowledge Is Power Program (KIPP) within Charter Schools in the USA." *Journal for Critical Education Policy Studies* 7 (2): 126–153.

Lacy, Karyn R. 2007. *Blue-Chip Black: Race, Class, and Status in the New Black Middle Class.* Berkeley: University of California Press.

Lafer, G. 2002. *The Job Training Charade.* Ithaca: Cornell University Press.

Lasar, Matthew. 2011. "There's a Mercedes Divide: Former FCC Chief Now Top Cable Lobbyist." *Ars Technica*, March 16, 2011. http://arstechnica.com/tech-policy/2011/03 /what-did-he-mean-by-that-mercedes-divide-fcc-chief-now-top-cable-lobbyist/.

Lashaw, A. 2008. "Experiencing Imminent Justice: The Presence of Hope in a Movement for Equitable Schooling." *Space and Culture* 11 (2): 109–124.

Lassiter, Luke E. 2005. *The Chicago Guide to Collaborative Ethnography*. Chicago: University of Chicago Press.

Lazere, Ed, and Marco Guzman. 2015. "Left Behind: DC's Economic Recovery Not Reaching All Residents." DC Fiscal Policy Institute, January 29, 2015. https://www .dcfpi.org/all/left-behind-dcs-economic-recovery-is-not-reaching-all-residents-2/.

Leman, Nicholas. 2013. "How Michelle Rhee Misled Education Reform." *New Republic*, May 20, 2013. https://newrepublic.com/article/113096/how-michelle-rhee -misled-education-reform.

Light, Jennifer S. 2001. "Rethinking the Digital Divide." *Harvard Educational Review* 71 (4): 709–734.

Lipman, Pauline. 2015. "Capitalizing on Crisis: Venture Philanthropy's Colonial Project to Remake Urban Education." *Critical Studies in Education* 56 (2): 241–258.

Liu, Yujia, and David B. Grusky. 2013. "The Payoff to Skill in the Third Industrial Revolution." *American Journal of Sociology* 118 (5): 1330–1374.

Loftus, Joseph A. 1967. "Commerce Agency Trains Youths on Computers." *New York Times*, August 13, 1967.

Lopez, Steven Henry. 2010. "Workers, Managers, and Customers: Triangles of Power in Work Communities." *Work and Occupations* 37 (3): 251–271.

Losen, Daniel J., Michael A. Keith, Cheri L. Hodson, and Tia E. Martinez. 2016. *Charter Schools, Civil Rights and School Discipline: A Comprehensive Review*. Los Angeles: The Center for Civil Rights Remedies at the Civil Rights Project, University of California, Los Angeles.

Losse, Kate. 2012. *The Boy Kings: A Journey into the Heart of the Social Network*. New York: Simon and Schuster.

Luyt, Brendan. 2001. "Regulating Readers: The Social Origins of the Readers' Advisor in the United States." *Library Quarterly* 71 (4): 443–466.

Marcus, George. 1995. "Ethnography in/of the World System: The Emergence of Multi-sited Ethnography." *Annual Review of Anthropology* 24:95–117.

Marx, Karl. 1990. *Capital: A Critique of Political Economy*. Vol. 1. London: Penguin.

Martell, Nevin. 2016. "Meet the Man Who Is Turning DC Libraries into a National Model." *Washington Post*, April 3, 2016. https://www.washingtonpost.com/lifestyle

/magazine/meet-the-man-who-is-turning-dcs-library-system-into-a-national-model /2016/03/30/5d06eda0-db50-11e5-891a-4ed04f4213e8_story.html.

Marwick, Alice E. 2013. *Status Update: Celebrity, Publicity, and Branding in the Social Media Age.* New Haven: Yale University Press.

McCartney, Robert. 2015. "Why DC Has a Uniquely Bad Record on Helping the Unemployed Get Jobs." *Washington Post,* December 12, 2015. https://www.washingtonpost .com/local/why-dc-has-a-uniquely-bad-record-on-helping-the-unemployed-get-jobs /2015/12/12/751d05ae-99d4-11e5-94f0-9eeaff906ef3_story.html.

McCartin, Joseph A. 2006. "Bringing the State's Workers In: Time to Rectify an Imbalanced US Labor Historiography." *Labor History* 47 (1): 73–94.

Mecanoo Architecten and Martinez + Johnson. 2014. MLK Modernization Final Designs. DC Public Library. https://www.slideshare.net/DCPublicLibrary/the-mlk-final -designs.

Medici, Andy. 2019. "Blackboard Is Leaving DC for Reston." *Washington Business Journal,* January 2, 2019. https://www.bizjournals.com/washington/news/2019/01/02 /this-education-technology-giant-is-leaving-d-c-for.html.

Miltner, Katharine M. 2019. "Anyone Can Code? The Coding Fetish and the Politics of Sociotechnical Belonging." PhD diss., University of Southern California.

Mohandesi, Salar, and Emma Teitelman. 2017. "Without Reserves." In *Social Reproduction Theory: Remapping Class, Recentering Oppression,* edited by Tithy Bhatttacharya, 37–67. London: Pluto Press.

Moser, Michele, and Ross Rubenstein. 2002. "The Equality of Public School District Funding in the United States: A National Status Report." *Public Administration Review* 62 (1): 63–72.

Murphy, Carlyle. 2004. "As Aged Building Breaks Down, Readership Is Up: Volunteers, Staff's Can-Do Spirit Help Ward 1 Library Meet New Challenges." *Washington Post,* September 30, 2004, sec. DE 08.

Nadasen, Premilla. 2002. "Expanding the Boundaries of the Women's Movement: Black Feminism and the Struggle for Welfare Rights." *Feminist Studies* 28 (2): 271–301.

Nakamura, Lisa. 2000. "Where Do You Want to Go Today? Cybernetic Tourism, the Internet, and Transnationality." In *Race in Cyberspace,* edited by Beth Kolko, Lisa Nakamura, and Gilbert Rodman, 15–26. New York: Routledge.

National Commission on Excellence in Education. 1983. *A Nation at Risk: The Imperative for Educational Reform.* Washington, DC: US Government Printing Office. http://www2.ed.gov/pubs/NatAtRisk/index.html.

Neff, Gina. 2012. *Venture Labor: Work and the Burden of Risk in Innovative Industries.* Cambridge, MA: MIT Press.

Neff, Gina, and David Stark. 2004. "Permanently Beta." In *Society Online: The Internet in Context*, edited by Philip N. Howard and Steve Jones, 173–188. Thousand Oaks: Sage Publications.

NSF (National Science Foundation). 2013. "Science and Engineering Degrees: 1966–2010." NSF 13-327. http://www.nsf.gov/statistics/nsf13327/content.cfm?pub_id=4266&id=2.

NTIA (National Telecommunications and Information Administration). 1995. *Falling through the Net: A Survey of "Have Nots" in Rural and Urban America*. Washington, DC: US Department of Commerce. http://www.ntia.doc.gov/ntiahome/fallingthru.html.

NTIA (National Telecommunications and Information Administration). 1998. *Falling through the Net II: New Data on the Digital Divide*. Washington, DC: US Department of Commerce. http://www.ntia.doc.gov/report/1998/falling-through-net-ii-new-data-digital-divide.

NTIA (National Telecommunications and Information Administration). 1999. *Falling through the Net III: Defining the Digital Divide*. Washington, DC: US Department of Commerce. http://www.ntia.doc.gov/legacy/ntiahome/fttn99/FTTN.pdf.

NTIA (National Telecommunications and Information Administration). 2000. *Falling through the Net IV: Toward Digital Inclusion*. Washington, DC: US Department of Commerce. http://www.ntia.doc.gov/files/ntia/publications/fttn00.pdf.

Nixon, Richard. 1969. "Address to the Nation on Domestic Programs." Weekly Compilation of Presidential Documents no. 5, 1103–1112.

Obama, Barack H. 2013. "President Obama Asks America to Learn Computer Science." Code.org, YouTube, December 3, 2013. https://www.youtube.com/watch?v=6XvmhE1J9PY.

Obama, Barack H. 2016. "Remarks by the President at the White House Science Fair." Office of the Press Secretary, April 13, 2016. https://www.whitehouse.gov/the-press-office/2016/04/13/remarks-president-white-house-science-fair.

O'Callaghan, Cian. 2010. "Let's Audit Bohemia: A Review of Richard Florida's 'Creative Class' Thesis and Its Impact on Urban Policy." *Geography Compass* 4 (11): 1606–1617.

ODMPED (Office of the Deputy Mayor for Planning and Economic Development). 2014. *Creative Economy Strategy for the District of Columbia*. Washington, DC: Office of the Deputy Mayor for Planning and Economic Development. https://dmped.dc.gov/publication/creative-economy-strategy-full-report.

OECD (Organization for Economic Cooperation and Development). 2019. "Broadband Portal." http://www.oecd.org/sti/broadband/broadband-statistics/.

Office of the Chief Financial Officer. 2004. "District Government Achieves Balanced Budget and Clean Audit Opinion for FY 2003." Press release, District of Columbia,

Office of the Chief Financial Officer, January 30, 2004. https://web.archive.org/web/20090508094540/http://app.cfo.dc.gov/CFORUI/news/release.asp?id=96&mon=200401.

Office of Revenue Analysis. 2017. *District of Columbia Economic and Revenue Trends: November 2017*. Washington, DC: District of Columbia Office of Revenue Analysis. https://cfo.dc.gov/sites/default/files/dc/sites/ocfo/publication/attachments/DC%20Economic%20and%20Revenue%20Trend%20Report%20_November%202017.pdf.

Olson, Gary M., and Jonathan Grudin. 2009. "The Information School Phenomenon." *Interactions* 16 (2): 15–19.

Osborne, David. 2015. *A Tale of Two Systems: Education Reform in Washington, DC*. Washington, DC: Progressive Policy Institute.

OSSE (Office of the State Superintendent of Education). 2013. *Reducing Out-of-School Suspensions and Expulsions in District of Columbia Public and Public Charter Schools*. Washington, DC: OSSE. http://osse.dc.gov/sites/default/files/dc/sites/osse/publication/attachments/OSSE_REPORT_DISCIPLINARY_G_PAGES.pdf.

OSSE (Office of the State Superintendent of Education). 2019. *District of Columbia Teacher Workforce Report*. Washington, DC: District of Columbia, OSSE. https://osse.dc.gov/sites/default/files/dc/sites/osse/publication/attachments/DC%20Educator%20Workforce%20Report%2010.2019.pdf.

Parrenas, Rhacel Salazar. 2012. "The Reproductive Labor of Migrant Workers." *Global Networks* 12 (2): 269–275.

Peck, Jamie. 2001. *Workfare States*. New York: Guilford Publications.

Peck, Jamie. 2005. "Struggling with the Creative Class." *International Journal of Urban and Regional Research* 29 (4): 740–770.

Peck, Jamie. 2007. "The Creativity Fix." *Eurozine*, June 28, 2007. https://www.eurozine.com/the-creativity-fix/.

Peck, Jamie. 2012. "Austerity Urbanism: American Cities under Extreme Economy." *City* 16 (6): 626–655.

Perrin, Andrew, and Maeve Duggan. 2015. "Americans' Internet Access: 2000–2015." Pew Research Center, June 26, 2015. https://www.pewresearch.org/internet/2015/06/26/americans-internet-access-2000-2015/.

Pew Research Center. 2014. "Public Libraries and Technology: From 'Houses of Knowledge' to 'Houses of Access.'" *Internet and Tech* (blog), Pew Research Center, July 9, 2014. https://www.pewresearch.org/internet/2014/07/09/public-libraries-and-technology-from-houses-of-knowledge-to-houses-of-access/.

Piven, Frances Fox. 1998. "Welfare and Work." *Social Justice* 25 (1): 67–81.

Piven, Frances Fox, and Richard Cloward. 2012. *Regulating the Poor: The Functions of Public Welfare*. New York: Vintage.

Platt, Eric, and Andrew Edgecliffe-Johnson. 2020. "WeWork: How the Ultimate Unicorn Lost Its Billions." *Financial Times*, February 20, 2020. https://www.ft.com/content/7938752a-52a7-11ea-90ad-25e377c0ee1f.

Press, Alex. 2019. "On the Origins of the Professional-Managerial Class: An Interview with Barbara Ehrenreich." *Dissent*, October 22, 2019. https://www.dissentmagazine.org/online_articles/on-the-origins-of-the-professional-managerial-class-an-interview-with-barbara-ehrenreich.

Ray, Victor. 2019. "A Theory of Racialized Organizations." *American Sociological Review* 84 (1): 26–53.

Reese, Ellen, and Garnett Newcombe. 2003. "Income Rights, Mothers' Rights, or Workers' Rights? Collective Action Frames, Organizational Ideologies, and the American Welfare Rights Movement." *Social Problems* 50 (2): 294–318.

Reid, Ian. 2017. "The 2017 Public Library Data Service Report: Characteristics and Trends." *Public Libraries Online*, December 4, 2017. http://publiclibrariesonline.org/2017/12/the-2017-public-library-data-service-report-characteristics-and-trends/.

Rein, Lisa. 2014. "As Federal Government Evolves, Its Clerical Workers Edge toward Extinction." *Washington Post*, January 14, 2014. http://www.washingtonpost.com/politics/as-federal-government-evolves-its-clerical-workers-edge-toward-extinction/2014/01/14/ded78036-5eae-11e3-be07-006c776266ed_story.html.

Reveley, James, and Simon Ville. 2010. "Enhancing Industry Association Theory: A Comparative Business History Contribution." *Journal of Management Studies* 47 (5): 837–858.

Ries, Eric. 2011. *The Lean Startup: How Today's Entrepreneurs Use Continuous Innovation to Create Radically Successful Businesses*. New York City: Crown Business.

Rios, Victor M. 2015. Review of *On the Run: Fugitive Life in an American City* by Alice Goffman. *American Journal of Sociology* 121 (1): 306–308.

Rivers, Wes. 2015. "Going, Going, Gone: DC's Vanishing Affordable Housing." DC Fiscal Policy Institute, March 12, 2015. https://www.dcfpi.org/all/going-going-gone-dcs-vanishing-affordable-housing-2/.

Robinson, Laura, Shelia R. Cotten, Hiroshi Ono, Anabel Quan-Haase, Gustavo Mesch, Wenhong Chen, Jeremy Schulz, Timothy M. Hale, and Michael J. Stern. 2015. "Digital Inequalities and Why They Matter." *Information, Communication & Society* 18 (5): 569–582.

Rogers, Everett M. 2010. *Diffusion of Innovations*. New York: Simon and Schuster.

Romm, Tony. 2017. "President Donald Trump and His Daughter Ivanka Are Unveiling a New Federal Computer Science Initiative with Major Tech Backers." Vox,

September 25, 2017. https://www.vox.com/2017/9/25/16276904/president-donald -trump-ivanka-tech-stem-computer-science-coding-education-amazon-google.

Ross, Andrew. 2004. *No-Collar: The Humane Workplace and Its Hidden Costs*. Philadel- phia: Temple University Press.

Roth, Erin, and Will Perkins. 2019. *DC Schools Shortchange At-Risk Students*. Washing- ton, DC: Office of the District of Columbia Auditor.

Ryan, Camille L., and Kurt Bauman. 2016. *Educational Attainment in the United States: 2015*. Open-file report P20-578. Washington, DC: United States Census Bureau. https://www.census.gov/content/dam/Census/library/publications/2016/demo/p20 -578.pdf.

Sassen, Saskia. 2001. *The Global City: New York, London, Tokyo*. Princeton: Princeton University Press.

Schaffner, Brian F., and Patrick J. Sellers, eds. 2009. *Winning with Words: The Origins and Impact of Political Framing*. New York: Routledge.

Schatteman, Alicia, and Ben Bingle. 2015. "Philanthropy Supporting Government: An Analysis of Local Library Funding." *Journal of Public and Nonprofit Affairs* 1 (2): 74–86.

Schwartz, John. 2002. "Report Disputes Bush Approach to Bridging 'Digital Divide.'" *New York Times*, July 11, 2002. https://www.nytimes.com/2002/07/11/us/report-dispu tes-bush-approach-to-bridging-digital-divide.html.

Schwartzman, Paul, and Chris L. Jenkins. 2010. "How DC Mayor Fenty Lost the Black Vote—and His Job." *Washington Post*, September 18, 2010. http://www.washingtonpost .com/wp-dyn/content/article/2010/09/18/AR2010091804286.html.

Scott, George A. 2012. *Charter Schools: Additional Federal Action Needed to Help Protect Access for Students with Disabilities*. Report to Congressional Requesters GAO-12-543. United States Government Accountability Office.

Scott, Janelle T. 2009. "The Politics of Venture Philanthropy in Charter School Policy and Advocacy." *Educational Policy* 23 (1): 106–136.

Scott, Janelle T. 2013. "A Rosa Parks moment? School choice and the marketization of civil rights." *Critical Studies in Education* 54 (1): 5–18.

Scott, Janelle T., Tina Trujillo, and Marialena D. Rivera. 2016. "Reframing Teach for America: A Conceptual Framework for the Next Generation of Scholarship." *Educa- tion Policy Analysis Archives* 24 (12). http://dx.doi.org/10.14507/epaa.24.2419.

Scott, W. Richard. 1991. "Unpacking Institutional Arguments." In *The New Institu- tionalism in Organizational Analysis*, edited by Walter W. Powell and Paul J. DiMag- gio, 164–182. Chicago: University of Chicago Press.

Seamster, Louise, and Raphaël Charron-Chénier. 2017. "Predatory Inclusion and Education Debt: Rethinking the Racial Wealth Gap." *Social Currents* 4 (3): 199–207.

Seaver, Nicholas. 2015. "Computing Taste: The Making of Algorithmic Music Recommendation." PhD diss., University of California, Irvine.

Shaefer, Luke, and Kathryn Edin. 2013. "Rising Extreme Poverty in the United States and the Response of Federal Means-Tested Transfer Programs." *Social Service Review* 87 (2): 250–268.

Shaefer, Luke, Kathryn Edin, and Elizabeth Talbert. 2015. "Understanding the Dynamics of $2-a-Day Poverty in the United States." *RSF: The Russel Sage Foundation Journal of the Social Sciences* 1 (1): 120–138.

Shedd, Carla. 2015. *Unequal City: Race, Schools, and Perceptions of Injustice*. New York: Russell Sage Foundation.

Sheir, Rebecca. 2015. "DC Public Library Expands Outreach to Homeless Patrons." WAMU 88.5, Metro Connection, February 20, 2015. http://wamu.org/programs/metro_connection/15/02/20/dc_public_library_expands_outreach_to_homeless_patrons.

Sherwood, Tom, and Harry S. Jaffe. 2014. *Dream City: Race, Power, and the Decline of Washington, DC*. New York: Simon & Schuster.

Silver, Beverly J. 2003. *Forces of Labor: Workers' Movements and Globalization since 1870*. Cambridge, MA: Cambridge University Press.

Simon, Stephanie. 2013. "Class Struggle: How Charter Schools Get Students They Want. *Reuters*, February 15, 2012. https://www.reuters.com/article/us-usa-charters-admissions/special-report-class-struggle-how-charter-schools-get-students-they-want-idUSBRE91E0HF20130215.

Sims, Calvin. 1992. "Silicon Valley Takes a Partisan Leap of Faith." *New York Times*, October 29, 1992. http://www.nytimes.com/1992/10/29/business/silicon-valley-takes-a-partisan-leap-of-faith.html.

Sims, Christo. 2017. *Disruptive Fixation: School Reform and the Pitfalls of Techno-Idealism*. Princeton: Princeton University Press.

Sirin, Selcuk R. 2005. "Socioeconomic Status and Academic Achievement: A Meta-analytic Review of Research." *Review of Educational Research* 75 (3): 417–453.

Smith, Andrew. 2015. *Searching for Work in the Digital Era*. Washington, DC: Pew Research Center. https://www.pewresearch.org/internet/2015/11/19/searching-for-work-in-the-digital-era/.

Smith, Dorothy E. 2005. *Institutional Ethnography: A Sociology for People*. Lanham: AltaMira Press.

Smith, Neil. 2002. "New Globalism, New Urbanism: Gentrification as Global Urban Strategy." *Antipode* 34 (3): 427–450.

Smith, Neil. 2005. *The New Urban Frontier: Gentrification and the Revanchist City*. New York: Routledge.

Smith, William D. 1968. "Companies Aid Computer School." *New York Times*, July 12, 1968, 39.

Smothers, Ronald. 1981. "CETA Cutbacks Leaving Thousands Unemployed." *New York Times*, April 11, 1981, 1.

Sojoyner, Damien M. 2013. "Black Radicals Make for Bad Citizens: Undoing the Myth of the School to Prison Pipeline." *Berkeley Review of Education* 4 (2). http://dx.doi.org/10.5070/B84110021.

Spence, Lester K. 2015. *Knocking the Hustle: Against the Neoliberal Turn in Black Politics*. New York: Punctum Books.

Stehlin, John. 2016. "The Post-industrial 'Shop Floor': Emerging Forms of Gentrification in San Francisco's Innovation Economy." *Antipode* 48 (2): 474–493.

Steinberg, Ronnie J. 1990. "Social Construction of Skill: Gender, Power, and Comparable Worth." *Work and Occupations* 17 (4): 449–482.

Stevenson, Siobhan A. 2009. "Digital Divide: A Discursive Move Away from the Real Inequities." *Information Society* 25 (1): 1–22.

Stevenson, Siobhan. 2010. "The Political Economy of Andrew Carnegie's Library philanthropy, with a Reflection on Its Relevance to the Philanthropic Work of Bill Gates." *Library & Information History* 26 (4): 237–257.

Stevenson, Siobhan A. 2011. "New Labor in Libraries: The Post-Fordist Public Library." *Journal of Documentation* 67 (5): 773–790.

Stevenson, Siobhan A., and Caleb Domsy. 2016. "Redeploying Public Librarians to the Front-Lines: Prioritizing Digital Inclusion." *Library Review* 65 (6–7): 370–385.

Straubhaar, Joseph, Jeremiah Spence, Karen Gustoffsen, Maria Rios, Fabio Ferreira, and Vanessa Higgins. 2008. "Comparative Analysis of Information Society Discourse and Public Policy Responses in the United States and Brazil." *Logos* 28 (15): 84–104.

Sturtevant, Lisa. 2014. "The New District of Columbia: What Population Growth and Demographic Change Mean for the City." *Journal of Urban Affairs* 36 (2): 276–299.

Sugrue, Thomas J. 2014. *The Origins of the Urban Crisis: Race and Inequality in Postwar Detroit*. Princeton: Princeton University Press.

TallBear, Kim. 2014. "Standing with and Speaking as Faith: A Feminist-Indigenous Approach to Inquiry." *Journal of Research Practice* 10 (2): article N17. http://jrp.icaap.org/index.php/jrp/article/view/405/371.

Tavernise, Sabrina. 2011. "A Population Changes, Uneasily." *New York Times*, July 17, 2011. http://www.nytimes.com/2011/07/18/us/18dc.html.

Taylor, Frank V. 2016. *2016 Startup Census Report: Greater Washington, DC Region*. Washington, DC: Fosterly. http://fosterly.com/wp-content/uploads/2017/10/2016-Fosterly-Census-Report-digital-x-small.pdf.

Teresa, Benjamin F., and Ryan M. Good. 2018. "Speculative Charter School Growth in the Case of UNO Charter School Network in Chicago." *Urban Affairs Review* 54 (6): 1107–1133.

Thompson, Edward P. 1991. *The Making of the English Working Class*. New York: Penguin Books.

Thornton, Patricia H., and William Ocasio. 1999. "Institutional Logics and the Historical Contingency of Power in Organizations: Executive Succession in the Higher Education Publishing Industry, 1958–1990." *American Journal of Sociology* 105 (3): 801–843.

Thornton, Patricia H., and William Ocasio. 2008. "Institutional Logics." In *The Sage Handbook of Organizational Institutionalism*, edited by Royston Greenwood, Christine Oliver, Kerstin Sahlin, and Roy Suddaby, 99–128. London: Sage Publications.

Tienken, Christopher H., Anthony Colella, Christian Angelillo, Meredith Fox, Kevin R. McCahill, and Adam Wolfe. 2017. "Predicting Middle Level State Standardized Test Results Using Family and Community Demographic Data." *RMLE Online* 40 (1): 1–13.

Tinbergen, Jan. 1974. "Substitution of Graduate by Other Labor." *Kyklos* 27: 217–226.

Tissot, Sylvie. 2015. *Good Neighbors: Gentrifying Diversity in Boston's South End*. New York: Verso Books.

Tochterman, Brian. 2012. "Theorizing Neoliberal Urban Development: A Genealogy from Richard Florida to Jane Jacobs." *Radical History Review* 112 (Winter 2012): 65–87.

Toppo, Greg. 2013. "Memo Warns of Rampant Cheating in DC Public Schools." *USA Today*, April 11, 2013. http://www.usatoday.com/story/news/nation/2013/04/11/memo-washington-dc-schools-cheating/2074473/.

Tracey, Paul, Nelson Phillips, and Owen Jarvis. 2011. "Bridging Institutional Entrepreneurship and the Creation of New Organizational Forms: A Multilevel Model." *Organization Science* 22 (1): 60–80.

Trisi, Danilo, and Matt Saenz. 2020. *Deep Poverty among Children Rose in TANF's First Decade, then Fell as Other Programs Strengthened: Stronger Income Support Policies Needed to Make Further Progress*. Washington, DC: Center on Budget and Policy Priorities. https://www.cbpp.org/research/poverty-and-inequality/deep-poverty-among-children-rose-in-tanfs-first-decade-then-fell-as.

Turque, Bill. 2010. "Foundations Reserve Right to Pull Funding if DC Schools Chief Rhee Leaves." *Washington Post*, April 28, 2010. http://www.washingtonpost.com/wp-dyn/content/article/2010/04/27/AR2010042702791.html.

Turque, Bill, and Jon Cohen. 2010. "DC Schools Chancellor Rhee's Approval Rating in Deep Slide." *Washington Post*, February 1, 2010. http://www.washingtonpost.com /wp-dyn/content/article/2010/01/31/AR2010013102757.html.

Uetricht, Micah. 2014. *Strike for America: Chicago Teachers against Austerity*. New York: Verso Books.

US Census Bureau. 2019. *Revised Fiscal Year 2017 Annual Survey of School System Finances*. Washington, DC: US Census Bureau.

Vaisey, Stephen. 2006. "Education and Its Discontents: Overqualification in America, 1972–2002." *Social Forces* 85 (2): 835–864.

van Dijk, Jan A. G. M. 2005. *The Deepening Divide: Inequality in the Information Society*. Thousand Oaks: Sage Publications.

Vasquez, Julian Heilig, Muhammad Khalifa, and Linda C. Tillman. 2014. "High-Stake Reforms and Urban Education." In *Handbook of Urban Education*, edited by H. Richard Miller and Kofi Lomotey, 523–538. New York: Routledge.

Viseu, Ana, Andrew Clement, Jane Aspinall, and Tracy L. M. Kennedy. 2006. "The Interplay of Public and Private Spaces in Internet Access." *Information, Community & Society* 9 (5): 633–656.

Vogel, Lise. 2013. *Marxism and the Oppression of Women: Toward a Unitary Theory*. Boston: Brill.

Wacquant, Loïc. 1999. "Urban Marginality in the Coming Millennium." *Urban Studies* 36 (10): 1639–1647.

Wacquant, Loïc. 2009. *Punishing the Poor: The Neoliberal Government of Social Insecurity*. Durham: Duke University Press.

Wacquant, Loïc. 2012. "Three Steps to a Historical Anthropology of Actually Existing Neoliberalism." *Social Anthropology* 20 (1): 66–79.

Warschauer, Mark. 2002. "Reconceptualizing the Digital Divide." *First Monday* 7 (1). http://firstmonday.org/article/view/967/888/.

Warschauer, Mark. 2004. *Technology and Social Inclusion: Rethinking the Digital Divide*. Cambridge, MA: MIT Press.

Watkins, S. Craig. 2018. *The Digital Edge: How Black and Latino Youth Navigate Digital Inequality*. New York: NYU Press.

Wayne, Leslie. 1982. "Designing a New Economics for the 'Atari Democrats.'" *New York Times*, September 26, 1982, F6.

Weber, Lauren. 2016. "Dropouts Need Not Apply: Silicon Valley Asks Mostly for Developers with Degrees." *Wall Street Journal*, March 30, 2016. http://blogs.wsj.com

/economics/2016/03/30/dropouts-need-not-apply-silicon-valley-asks-mostly-for
-developers-with-degrees/.

Weber, Rachel. 2002. "Extracting Value from the City: Neoliberalism and Urban
Redevelopment." *Antipode* 34 (3): 519–540.

Wiegand, Wayne. 1986. *The Politics of an Emerging Profession: The American Library
Association 1876–1917*. Westport: Greenwood.

Wiggins, Andrea, and Steve Sawyer. 2012. "Intellectual Diversity and the Faculty
Composition of iSchools." *Journal of the American Society for Information Science and
Technology* 63 (1): 8–21.

Wilhelm, Anthony G. 2003. "Leveraging Sunken Investments in Communications
Infrastructure: A Policy Perspective from the United States." *Information Society* 19 (4):
279–286.

Williams, Anthony A. 2005. "Draft Technical Report of the Mayor's Task Force on
the Future of the District of Columbia Public Library System." November 2005.

Williams, Brett. 1988. *Upscaling Downtown: Stalled Gentrification in Washington*. Ithaca:
Cornell University Press.

Williams, Raymond. 1973. "Base and Superstructure in Marxist Cultural Theory."
New Left Review 82 (November/December): 3–16.

Williamson, Ben. 2016. "Political Computational Thinking: Policy Networks, Digital
Governance and 'Learning to Code.'" *Critical Policy Studies* 10 (1): 39–58.

Willis, Paul E. 1981. *Learning to Labor: How Working Class Kids Get Working Class Jobs*.
New York: Columbia University Press.

Willse, Craig. 2015. *The Value of Homelessness: Managing Surplus Life in the United States*.
Minneapolis: University of Minnesota Press.

Wilson, Valerie. 2015. "Recovery of Hispanic Unemployment Rate Expands to Four
More States in Third Quarter of 2015." Economic Policy Institute, November 3, 2015.
https://www.epi.org/publication/recovery-of-hispanic-unemployment-rate-expands
-to-four-more-states-in-third-quarter-of-2015/.

Winant, Gabriel. 2019. "Professional-Managerial Chasm: A Sociological Designation
Turned into an Epithet and Hurled like a Missile." *N+1 Magazine*, October 10, 2019.
https://nplusonemag.com/online-only/online-only/professional-managerial-chasm/.

Windham, Lane. 2017. *Knocking on Labor's Door: Union Organizing in the 1970s and
the Roots of a New Economic Divide*. Chapel Hill: University of North Carolina Press.

Winner, Langdon. 1980. "Do Artifacts Have Politics?" *Daedalus* 109 (1): 121–136.

Wolch, Jennifer. 2007. "Green Urban Worlds." *Annals of the Association of American
Geographers* 97 (2): 373–384.

Zillien, Nicole, and Eszter Hargittai. 2009. "Digital Distinction: Status-Specific Types of Internet Usage." *Social Science Quarterly* 90 (2): 274–291.

Zuckerman, Adam, Frank V. Taylor, Harry Alford III, Pam Rothenberg, Ryan Touhill, and Theo Slagle. 2017. *2017 Startup Census Report: Greater Washington, DC Region.* Washington, DC: Fosterly. http://fosterly.com/wp-content/uploads/2018/02/2017 -Fosterly-Census-Report-Small.pdf.

Index

1776 (startup incubator), 29, 64–65, 72, 75–77, 184
2U (startup), 142

Academic culture
  discipline and, 112, 114–116, 121, 132–138
  Du Bois and, 112–116, 121–133, 136–138
  hidden curriculum and, 121, 123–128, 131–132, 135, 138, 150
  high-performance, 112–116, 121–133, 136–138
  presence bleed and, 127–133
  student computers and, 122–123
  technology and, 121–133
Access doctrine
  bootstrapping and, 15–16, 24, 30, 61, 85, 101, 143–145, 148, 150, 153, 160, 166–170, 173, 177, 180, 184–185, 189, 201n3
  charter schools and, 112, 115–117, 120
  Clinton and, 30–31, 38–39, 48, 53, 55, 57, 60, 144, 150, 174
  competing institutional cultures and, 165–170
  digital divide and, 1–2, 6–13, 16, 24, 30–31, 36, 38–39, 43, 48, 57, 60, 90, 116, 144–145, 170
  Du Bois and, 116

free Wi-Fi and, 2, 186
  future of, 52–58
  ideal workers and, 181–186
  labor market and, 11, 13, 16, 18, 23, 32–33, 38, 43, 96, 144, 170, 177, 201n3
  libraries and, 25, 84–85, 89–90, 95–97, 101, 189
  MLK and, 16, 116
  new economy and, 24, 30–33, 36–39, 43–49, 53, 55, 57, 173–174
  Obama and, 11
  pivots and, 60–61, 66, 75
  political economy and, 5, 14, 22, 38, 49, 60, 144–145, 170, 174
  poverty and, 5, 7, 11–16, 24, 30–32, 38, 43, 46, 49, 53, 55, 57, 60–61, 75, 85, 90, 120, 143–144, 160, 166, 169, 174, 180
  predatory inclusion and, 169
  problem of, 173–177
  research methodology on, 26–28
  technology policy and, 31–38
  Washington, DC, and, 7, 18–26, 30, 32, 53, 57, 75, 95, 150, 184–185
Acer Chromebooks, 122
Activists
  bootstrapping and, 147
  Cameron, 173, 188
  charter schools and, 117
  education and, 117, 199n8

Activists (cont.)
  homeless and, 3, 147, 173, 197n1
  King and, 82
  new economy and, 34, 38, 49
ACT scores, 134
Acumen Solutions, 112, 199n1
Adams Place Day Center, 172
Adobe Creative Suite, 93, 95
Advanced Placement (AP) courses,
    112–114, 127, 134–135, 163
Advisory Neighborhood Commissions
    (ANCs), 88, 197n4
Agenda for Action (NII), 49
Aid for Families with Dependent
    Children (AFDC), 35, 37, 44
American Federation of State, County
    and Municipal Employees (AFSCME),
    154–155
American Library Association, 87
Android, 134
AOL, 64
AP for All, 163
Apple, 27, 41, 145
Armstrong, Richard, 159
AT&T, 50–51, 53
Atari Democrats
  bootstrapping and, 145, 151, 183
  neoliberalism and, 24, 30, 37, 40–43,
    52, 145, 151, 183
  new economy and, 30, 37, 40–42, 52
Austerity, 5, 13–14, 81, 95, 147, 165, 183
Automation, 31, 33, 95

Baltimore Inner Harbor, 22
Bankruptcy, 22, 181
Bekelman, Alan, 33
Belgium, 53
Bell, Terrel, 42
Bike and Roll, 103
Bill & Melinda Gates Foundation, 95,
    118, 143, 147, 154, 157–158, 162–164
Blackboard, 77, 79
Black Lives Matter movement, 127, 185

Black people
  access doctrine and, 1–2, 193n2
  bootstrapping and, 142–143, 147, 150,
    152–156, 168, 201nn2,3, 202n5
  charter schools and, 112–113, 117–
    119, 125, 127, 138, 199n7, 200n13
  crime and, 76, 171, 183, 191
  entrepreneurs and, 103, 117, 143, 150
  gentrification and, 18, 76, 172, 185
  Harris campaign and, 4
  libraries and, 185
  middle class and, 1, 20–21, 33, 76–77,
    82, 93, 119
  new economy and, 31–36, 46–47, 53
  police and, 22, 103, 171
  poverty and, 31–33, 53, 142–143, 175,
    183
  professionals and, 18–19, 87, 89, 93,
    119, 125, 138, 143, 150, 183, 185
  unemployment and, 6, 20, 29, 32–33
  wage increases and, 20
  Washington, DC, neighborhoods and,
    18–20
  welfare and, 32, 35–36, 185, 193n4
  women and, 17, 21, 142–143, 147, 153
Bloomberg, Michael, 64
Booms, economic, 19, 34, 64–65, 76, 81,
    193n4, 194n6
Bootcamps, 64, 66, 72, 75
Bootstrapping
  access doctrine and, 15–16, 24, 30,
    61, 85, 101, 143–145, 148, 150, 153,
    160, 166–170, 173, 177, 180,
    184–185, 189, 201n3
  activists and, 147
  Atari Democrats and, 145, 151, 183
  Black people and, 142–143, 147, 150,
    152–156, 168, 201nn2,3, 202n5
  budgets and, 162, 164
  capitalism and, 16, 167, 170, 177, 184
  carceral state and, 165, 169–170
  charter schools and, 24–26, 112,
    115, 119–120, 126, 137, 139–142,

145–148, 152, 155, 158, 160,
164–168, 175, 184, 194n10
CityBridge Foundation and, 141,
143–144, 148, 153–154, 156–157,
200n12, 201n1
Clinton and, 144, 150–151, 167,
203n1
competition and, 165–170, 174
computers and, 141, 155, 159–161,
164
crime and, 203n1
digital divide and, 16, 24, 30, 58, 90,
144–145, 164, 170
education and, 141–142, 147–153,
156–169, 201nn1–3, 202n6, 203n8
entrepreneurs and, 141, 143, 145,
149–150, 154, 157–161, 163, 201n1
failure of, 177–178
gentrification and, 159
Gore and, 150–151
homeless and, 146–147, 159–160
human capital and, 143, 150, 153,
157, 159–160, 165, 170, 191, 201n1
InCrowd and, 24–25, 61, 79, 115,
145–146, 165, 184, 194n10
inequality and, 162, 169
institutional culture of, 15–16
internet and, 161, 164, 175
labor market and, 144, 150, 159, 165,
177
Latinx and, 147, 156, 168
libraries and, 15–16, 24–26, 30, 79, 85,
89–90, 95, 97–108, 115, 120, 126,
144–155, 158–169, 184, 189, 194n10
meritocracy and, 148–153, 181,
202n6, 203n7
middle class and, 149
migrants and, 149
mission ambiguity and, 160–165
mobility and, 25, 158, 169, 175
neoliberalism and, 21, 24, 144,
154–155, 160, 165, 169, 173, 177,
184, 191, 201nn2,3

NewSchools Venture Fund and, 141,
147, 153, 157–158, 201n1
Obama and, 142–143
pathways of, 148
philanthropies and, 141, 145,
157–158, 162–165, 201n1
pivots and, 61, 79
police and, 191
poverty and, 15–16, 24, 30, 61, 85,
142–148, 153, 160–161, 165–166,
169, 175, 194, 201n1
predatory inclusion and, 169
professionals and, 143–145, 148–166,
169, 203n7
punishment and, 165
recession and, 164, 167, 179, 183
recruitment and, 156
reform and, 141, 146, 148, 151, 154,
156–158
regulation and, 145, 165
resisting, 100–105
revenue and, 15, 145, 164–165, 169
segregation and, 160, 162, 168, 203n1
skills and, 13, 15, 18, 24, 30, 61, 85,
106–107, 115, 141–144, 148–152,
159, 161, 166, 168, 170, 175, 177,
202n6
social reproduction and, 144, 146,
166–169, 173, 177, 180
software and, 164–165, 174, 176, 182,
184, 202n6
startups and, 15, 24–25, 58, 61,
115, 142–149, 153–154, 159–162,
165–166, 169, 175–176, 180, 184,
194nn9,10, 203n9
state disinvestment and, 18
taxes and, 145, 151, 162, 168
Teach for America (TFA) and, 155–158,
168, 188
technological professionalization and,
153–160
urban development and, 21–22
violence and, 159, 167–168, 170

Bootstrapping (cont.)
  welfare and, 144, 159, 165, 167,
    169–170, 176, 201n3, 203n1
  White people and, 146, 149–153,
    156–157, 161, 169
  women and, 142–143, 147, 153,
    202n5
Bowser, Muriel, 78–79, 141
Brazil, 54
Bridging organizations, 63, 75, 144,
    147, 153–154, 161, 166, 175, 195n2
"Bridging the Digital Divide in the
    District of Columbia" symposium, 29
Broad Academy, 157–158
Broadband, 4, 14, 29, 52–53, 133,
    195nn4,5
Budgets, 1
  austerity, 5, 13–14, 81, 95, 147, 165,
    183
  bootstrapping and, 162, 164
  libraries and, 14, 81, 95–96, 162, 164
  new economy and, 35–40
  philanthropies and, 14, 164
  recession and, 19, 37, 164
  schools and, 14, 95, 113, 162, 187,
    191
  Washington, DC, and, 19, 81
Bureau of Labor Statistics, 9, 151, 195n3
Burfield, Evan, 64
Bush, George W., 4, 8, 35–36, 39, 41, 56
Business-to-business (B2B) companies,
    59, 74–75
Business-to-consumer (B2C) companies,
    59

Cable, 32, 51
Canada, 88–89, 95
Capitalism
  bootstrapping and, 16, 167, 170, 177,
    184
  Carnegie and, 182
  charter schools and, 182
  competition and, 178

entrepreneurs and, 17, 22–23, 57, 62,
    78, 117, 143, 157, 159, 182, 200n12,
    201n1
  ideal workers and, 181–186
  inequality and, 5, 10, 13–14, 27–28,
    31, 43, 47–48, 54–55, 125, 162, 169,
    191, 193n2
  Keynesianism and, 23, 31–32, 36, 39,
    41, 44, 50, 55, 193
  libraries and, 180, 197n1
  new economy and, 40, 42, 195n3
  pivots and, 60, 76
  Rockefeller and, 182
  social divisions and, 5–6, 17, 178,
    183–184, 188
  social reproduction and, 16–17,
    177–182, 184, 195n3
  startups and, 17, 30, 62, 78, 143, 146,
    159, 175, 180–181, 184, 188
  violence and, 17
Capital One Arena, 22
Cappelli, Peter H., 10
Caputo-Pearl, Alex, 187
Carceral state
  bootstrapping and, 165, 169–170
  expansion of, 31, 60–61, 165
  labor market and, 34
  new economy and, 31–32, 34, 37–40,
    48, 57
  political economy and, 183
  poor neighborhoods and, 31, 183,
    203n1
  prisons and, 34, 37–40, 48, 179, 183,
    191, 203n1
  welfare and, 60, 169–170, 198n10,
    203n1
Cardoso, Fernando, 54
Carnegie, Andrew, 147, 163–164, 182,
    198n8
Census Bureau, 45, 199n8
Center for Education Reform, 117
Certification, 1–2, 7, 12, 151
Challenge Cup, 77

Chan Zuckerberg Initiative, 154, 158
Charter schools
  academic culture and, 112–116,
    121–133, 136–138
  access doctrine and, 112, 115–117,
    120
  activists and, 117
  as beacon of hope, 116–121
  Black people and, 112–113, 117–119,
    125, 127, 138, 199n7, 200n13
  bootstrapping and, 24–26, 112, 115,
    119–120, 126, 137–142, 145–148,
    152, 155, 158, 160, 164–168, 175,
    184, 194n10
  capitalism and, 182
  Clinton and, 120
  competition and, 117, 165–170
  computers and, 122–127, 133
  DC-CAS test and, 114, 140
  DC Public Charter School Board
    (DCPCSB) and, 112, 116–117, 147
  digital divide and, 116, 188
  discipline and, 112, 114–116, 121,
    132–138
  Du Bois and, 111 (*see also* W. E. B. Du
    Bois Public Charter High School)
  entrepreneurs and, 25, 117, 200n12
  experiments of, 111, 114–121, 132,
    138–140, 201n14
  gentrification and, 16, 115, 129, 137
  Gore and, 120
  hidden curriculum and, 121, 123–128,
    131–132, 135, 138, 150
  homeless and, 115, 131
  Houston, 127–128
  human capital and, 112, 116–117,
    120, 135, 140, 175
  inequality and, 125
  internet and, 126, 129
  labor market and, 115, 201n3
  Latinx people and, 112–113, 117, 119,
    125, 138, 185
  meritocracy and, 149–153
  middle class and, 119
  mission ambiguity and, 160–165
  mobility and, 119
  neoliberalism and, 117, 139, 186–187,
    200n9
  New Orleans, 142, 200nn9,13
  philanthropies and, 112, 118
  police and, 115
  poverty and, 118, 120
  presence bleed and, 127–133
  professionals and, 119–120, 124–126,
    128, 131, 134–139
  recession and, 118, 129
  Reconceptualizing Early Childhood
    Education (RECE) and, 116, 130,
    135, 140
  reform and, 117–118, 120
  regulation and, 118, 125, 139–140
  research methodology on, 26–28
  skills and, 115–116, 120, 131
  software and, 112, 122, 200n11
  startups and, 25, 115, 119, 139
  STEM subjects and, 117, 119
  teacher demographics and, 202n5
  technological professionalization and,
    153–160
  violence and, 115–116, 128, 178
  Waiting for "Superman" and, 118
  welfare and, 117
  White people and, 112–113, 118,
    124–125, 138, 140
  women and, 18
Chasen, Michael, 77
Chavous, Kevin, 117
Chicago Teachers Union (CTU), 187
Childcare, 36, 178, 186
Children
  Aid for Families with Dependent
    Children (AFDC) and, 35
  bootstrapping and, 201n1
  charter schools and, 116–118, 128
    (*see also* Charter schools)
  elite, 77

Children (cont.)
 future and, 4
 labor market and, 17
 libraries and, 82, 186, 198n5
 new economy and, 174
 poverty and, 16, 35, 37, 118, 201n1
 status and, 184
Children's Internet Protection Act,
  198n5
China, 187
Chocolate City, 20, 76, 118
CityBridge Foundation, 141–144, 148,
  153–157, 200n12, 201n1
Clinton, Bill
 access doctrine and, 30–31, 38–39, 48,
   53, 55, 57, 60, 144, 150, 174
 bootstrapping and, 144, 150–151, 167,
   203n1
 charter schools and, 120
 Democratic Leadership Council (DLC)
   and, 36
 digital divide and, 24, 30–31, 36,
   38–39, 45, 48–49, 56, 60, 144
 Economic Report of the President and,
   43
 education and, 24, 39, 42–43, 45, 174,
   203n1
 Falling through the Net report and, 45
 Gore and, 4, 11, 29, 36, 38–46, 49–50,
   53, 60, 84, 120, 150–151
 human capital and, 43, 45
 Keynesianism and, 36, 39, 50, 55
 labor market and, 34, 39, 41, 43, 144,
   195n3, 203n1
 libraries and, 87
 National Education Summit and, 42
 neoliberalism and, 24, 36, 51, 54, 56,
   87, 144, 174, 203n1
 new economy and, 30–31, 34, 36–45,
   48–57, 174, 195n3
 NTIA and, 45–46
 pivots and, 60, 77
 prison and, 39–40, 203n1

 private schooling of, 77
 skills training and, 24
 State of the Union and, 55, 174
 taxes and, 4, 151
 technology and, 4, 30, 87
 unions and, 151
 welfare and, 34, 37, 42, 55, 60, 87,
   174, 203n1
Code for DC, 155
Coding, 4–5, 9, 11–12, 40, 68, 75, 155,
  179, 188
Cold War, 39–40, 45
College of Information Studies, 94,
  155
Communism, 38–40
Competition
 bootstrapping and, 165, 174
 capitalism and, 178
 charter schools and, 117, 165–170
 digital divide and, 32, 52, 57, 174–175
 entrepreneurs and, 179
 hypercompetition and, 9
 libraries and, 89, 165–170, 198n8
 new economy and, 32, 40, 46–57,
   174–175
 pivots and, 64, 77
Comprehensive Employment and
  Training Act (CETA), 8
Computers
 bootstrapping and, 141, 155, 159–161,
   164
 charter schools and, 122–127, 133
 coding and, 4–5, 9, 11–12, 40, 68, 75,
   155, 179, 188
 Digital Commons and, 12, 82–86,
   89–109, 146, 159–160, 171–173, 176,
   198n6, 203n8
 Dream Lab and, 82–83, 89, 93, 95,
   101–103, 106, 108–109, 111, 147,
   150, 161, 171, 189
 education and, 25, 111, 114–115,
   119–130, 133–135, 138, 181
 Fab Lab and, 83–84, 89, 102, 108–109

impact of, 1–5, 11–12
libraries and, 12, 25–26, 81–109, 146,
    159–160, 171–173, 176, 185,
    189–190, 198nn6,7, 203n8
new economy and, 30–33, 40, 45–46,
    48
as professional tools, 14
skills and, 11 (*see also* Skills)
startups and, 65, 73
treatment of in schools, 122
women and, 196n3
Computer science, 4, 11, 30, 155,
    196n3
Computer Science Education Week, 4
Congressional Research Service, 50
Connect.DC, 1, 7, 29–30, 53
Conservatism, 3, 36, 42, 183
Cooper, Ginnie, 96, 189
Council on Competitiveness, 41
Crawford, Susan, 51
Creative Economy Strategy, 76
Crime
    Black people and, 76, 171, 183, 191
    bootstrapping and, 203n1
    Clinton on, 174
    drugs and, 57, 86, 99, 102, 126, 183
    Gray on, 76
    moral panic over, 193n4
    new economy and, 37, 39, 174
    police and, 191 (*see also* Police)
    rates of, 193n4, 203n1
    social issues behind, 191
    stolen computers and, 122
    Violent Crime and Law Enforcement
        Act and, 37, 39
    war on, 183
    Washington, DC, and, 76, 191
Crystal City, 65
Current Population Survey, 8, 45

DC-CAS test, 114, 140
DC Public Charter School Board
    (DCPCSB), 112, 116–117, 147

DC Public Library (DCPL) system, 197n3
    branch roles of, 88
    computer virus of, 81
    downsizing, 81
    education and, 81–83, 94
    as essential, 185
    financial services and, 188
    Gates Foundation and, 143
    human capital and, 177
    labor market and, 81
    mission ambiguity and, 162
    MLK, 82 (*see also* Martin Luther
        King Jr. Memorial Library—Central
        Library)
    police and, 2, 82, 91, 98, 102, 104, 189
    poor condition of, 82
    reorganization of, 86, 88
    research in, 4–5
    Reyes-Gavilan and, 29–30, 86, 88, 94,
        96, 150, 189, 197n4, 198n11
    Stovall and, 188–189
    Task Force on the Future of the DC
        Public Library System and, 82
DC Public Schools (DCPS) system, 116,
    134, 139, 200n12, 201n1
DC Tech (promotional network), 77, 83,
    161–162, 181
DC tech (sector), 64–65, 68, 70, 72,
    75–79, 83, 161–162, 181, 184
Degrees, 6, 9, 86, 94, 149, 151–156, 169,
    197n2, 198n7, 201n3
Dell computers, 84, 105, 122
Democracy, 36, 40, 49
Democratic Leadership Council (DLC),
    36, 150–151
Democratic Socialists of America, 184
Democrats, 197n7
    bootstrapping and, 145, 150–151,
        183–184
    education and, 117
    neoliberalism and, 24, 30, 37, 40–43,
        52, 145, 151, 183
    new economy and, 30, 36–37, 40–43, 52

Democrats (cont.)
  Keynesianism and, 36
  pivots and, 61, 78
Democrats for Education Reform, 117
Department of Commerce, 33, 41, 45,
  49, 56
Department of Education, 147
Department of Labor, 8, 196n6
Deregulation, 31, 34, 38, 41, 48, 51–52,
  56, 165
Digital Commons, 12
Digital DC, 76–77, 181
Digital divide
  access doctrine and, 1–2, 6–13, 16, 24,
    30–31, 36, 38–39, 43, 48, 57, 60, 90,
    116, 144–145, 170
  bootstrapping and, 16, 24, 30, 58, 90,
    144–145, 164, 170
  "Bridging the Digital Divide in the
    District of Columbia" and, 29
  charter schools and, 116, 188
  Clinton and, 24, 30–31, 36, 38–39, 45,
    48–49, 56, 60, 144
  coding and, 4–5, 9, 11–12, 40, 68, 75,
    155, 179, 188
  competition and, 32, 52, 57, 174–175
  discovery of, 36–37
  future of access and, 52–58
  gaps in, 6–13
  gender and, 32
  haves/have-nots and, 11, 29, 48, 50,
    61, 84
  libraries and, 84, 90, 188
  market solutions of, 48–52
  as measurement program, 42–48
  migrants and, 23
  as national economic crisis, 38–42
  new economy and, 29–32, 36, 38–52,
    56–58, 180
  pivots and, 25, 60, 63, 79, 84
  poverty and, 11–13, 24, 30–32, 38–39,
    43, 47–48, 90, 144, 174–175
  predatory inclusion and, 169

  professionals and, 12, 90, 174, 188
  "right" side of, 25, 58, 60, 84, 175
  skills and, 1, 6–13, 15, 24–25, 30–31,
    38, 43, 47, 52, 60, 90, 116, 144, 170,
    174–176
Digital inclusion, 29–30, 38, 144
Diplomas, 20, 87, 113
Disabilities, 14, 32, 34, 115, 199n7,
  203n1
Disruption Corp, 65
Dole, Bob, 40
Domsy, Caleb, 95
Dot-com boom, 62, 65
Douglas Development Corp., 77
Downtown DC Business Improvement
  District, 84
Drabinski, Emily, 187–188
Dropouts, 152, 179, 202n6, 203n8
Drugs, 57, 86, 99, 102, 126, 183
Du Bois. See W. E. B. Du Bois Public
  Charter High School

Education
  activists and, 117, 199n8
  ACT scores and, 134
  adult, 39, 186, 188
  AP courses and, 112–114, 127,
    134–135, 163
  bootstrapping and, 141–142, 147–153,
    156–169, 201nn1–3, 202n6, 203n8
  bootstrapping and, 15–16 (see also
    Bootstrapping)
  Broad Academy and, 157–158
  certification and, 1–2, 7, 12, 151
  charter schools and, 116 (see also
    Charter schools)
  CityBridge Foundation and, 141,
    143–144, 148, 153–154, 156–157,
    200n12, 201n1
  Clinton and, 24, 39, 42–43, 45, 174,
    203n1
  coding and, 4–5, 9, 11–12, 40, 68, 75,
    155, 179, 188

competing institutional cultures and, 165–170

computers and, 25, 111, 114–115, 119–130, 133–135, 138, 181

costs of higher, 31

DC-CAS test and, 114, 140

DC Public Library (DCPL) system and, 81–83, 94

degrees and, 6, 9, 86, 94–96, 149, 151–156, 169, 197n2, 198n7, 201n3

Democrats for Education Reform and, 117

diplomas and, 20, 87, 113

discipline and, 112, 114–116, 121, 132–138

dropouts and, 152, 179, 202n6, 203n8

Du Bois and, 25 (*see also* W. E. B. Du Bois Public Charter High School)

entrepreneurs and, 62, 117, 141, 143, 154, 157–159, 163, 179–182, 201n1

expansion of, 20

funding and, 14, 16, 34, 51, 53–54, 113, 116–117, 139–141, 143, 146–147, 149, 153, 156–164, 179, 199n8, 200n12, 201n1

Gates on, 162–163

GED, 2, 107

Google for Education and, 154, 158

gospel of, 143–144, 153, 201n3

GPAs and, 111, 113–114, 119, 138

Grow with Google and, 158

H-1B visas and, 9

homework and, 27, 113, 125–129, 195n5, 199n1

job training and, 2, 8, 11, 32, 35, 66, 76, 161, 177, 196n6

libraries and, 11–12, 94

mass firings of teachers and, 118–119

master of library science (MLS) degree and, 94–96, 155, 198n7

meritocracy and, 149–153

mission ambiguity and, 160–165

mobility and, 15, 32, 119, 158, 162, 169, 175

National Education Summit and, 42

Nation at Risk report and, 42

neoliberalism and, 117, 165, 169, 174–175, 186, 201n2, 203n1

new economy and, 31–32, 37, 39, 42–46, 54

NewSchools Venture Fund and, 141, 147, 153, 157–158, 201n1

Obama and, 3–4, 9, 117–118, 142

Pell Grants and, 37, 48

phones and, 4, 27, 113–115, 120–130, 133–138, 142, 161, 167

pivots and, 62, 69

presence bleed and, 127–133

prison, 37

professionals and, 7, 27, 41–42, 62, 100, 125, 128, 139, 145, 148, 150, 153, 156, 158–159, 162, 164, 175, 182–183, 186–187

public, 117, 159, 163, 199n1, 203n1

public rights and, 17

recession and, 14

reform and, 117–118, 120, 141, 154–158, 175, 186, 200n13, 201nn1,2, 203n1

role modeling and, 66, 121, 123

SAT scores and, 42, 114, 134, 140

skills and, 15 (*see also* Skills)

social reproduction and, 16–18, 21, 26, 34, 144, 146, 169, 179, 181, 186–187, 190, 203n1

special, 116

standardized testing and, 42–43, 111, 117, 122, 127, 134, 140, 153, 199n8, 203n1

STEM, 4, 9, 15, 77, 83, 117, 119, 165, 176

taxpayer revolts and, 14

teacher demographics and, 202n5

Teach for America (TFA) and, 155–158, 168, 188

Education (cont.)
  Work Hard Grades and, 136–137
Education Innovation Summit, 141,
    143, 145, 153, 156, 162, 164, 180
Ehrenreich, Barbara, 182
Ehrenreich, John, 182
Eisenhower, Dwight D., 39
E. L. Haynes Public Charter High
    School, 141
Eli and Edythe Broad Education
    Foundation, 157–158
Elites, 41, 75, 77, 128, 202n6, 203n1
English, 40, 116, 188, 199n6
Entrepreneurial urbanism, 22–23, 57,
    179
Entrepreneurs
  Black, 103, 117, 143, 150
  bootstrapping and, 141, 143, 145,
    149–150, 154, 157–161, 163, 201n1
  Broad Academy and, 157–158
  capital and, 17, 22–23, 57, 62, 78, 117,
    143, 157, 159, 182, 200n12, 201n1
  charter schools and, 25, 117, 200n12
  competition and, 179
  digital professionals and, 182
  education and, 62, 117, 141, 143, 154,
    157–159, 163, 179–182, 201n1
  gentrification and, 22
  Harvey on, 23
  ideal, 179–180
  internet and, 25, 58, 62, 179
  as leaders, 176
  libraries and, 25, 81–83, 87, 89, 93,
    96–97, 101–103, 106–107
  new economy and, 31, 57–58
  pivots and, 59, 61–68, 70, 71, 76–79,
    196n5
  recruitment and, 22, 64–65
  as roaming autodidact and, 179–180
  social reproduction and, 180
  startups and, 16–17, 25, 58, 61–62, 65,
    68, 76, 78, 143, 149, 154, 159, 176,
    179

  urban centers and, 23–24
  White, 22–23, 25, 64, 149–150, 157,
    179
Eubanks, Virginia, 12–13

Fab Lab, 83–84, 89, 102, 108–109
Facebook, 2, 70, 102, 158, 196n3
Falling through the Net report, 45, 47, 50
Federal Communications Commission
    (FCC), 50–53, 56, 195n4
Feedback, 27, 68, 74, 132
Feminism, 16, 199n2, 203n10
Fenty, Adrian, 163
Financial Control Board, 81
Financial crisis of 2008, 6, 14, 18–21,
    29, 56, 84, 95, 183
Food stamps, 92, 97, 203n1
Ford, Henry, 180
Ford Foundation, 158
Fort, The, 65
Fraser, Nancy, 187
Free trade, 36, 51
Free Wi-Fi, 2, 186
Friends of the Library, 82, 103, 107, 173,
    186, 189, 198n11
Funding
  education and, 14, 16, 34, 51, 53–54,
    113, 116–117, 139–141, 143,
    146–147, 149, 153, 156–164, 179,
    199n8, 200n12, 201n1
  libraries and, 14, 34, 51, 53–54, 81–83,
    90, 94–95, 146–147, 149, 153, 161,
    164–165, 179, 198nn5,8
  per-pupil, 140, 199n8, 200n12
  philanthropies and, 156 (see also
    Philanthropies)
  Race to the Top and, 147
  seed money and, 60

Gallery Place, 22
Gates, Bill, 95, 118, 143, 147, 154,
    157–158, 162–164, 202n6
GED, 2, 107

Gender
  digital divide and, 32
  libraries and, 87
  professional class and, 150
  skills and, 8, 32
  wage gaps and, 6, 8, 35, 69
General Assembly, 64
Gentrification
  Black people and, 18, 76, 172, 185
  bootstrapping and, 159
  charter schools and, 16, 115, 129, 137
  conflict over, 159, 186
  countergentrification research and, 185
  entrepreneurs and, 22
  libraries and, 81, 108, 172
  new economy and, 57
  pivots and, 63, 76, 78
  postrecession, 129
  startups and, 63, 76, 78, 181
  violence and, 181
  Washington, DC, and, 18, 63, 76, 78,
    81, 108, 172, 181, 197n4
  White people and, 22, 63, 76, 181
George, Sajan, 142–143, 154
Gephardt, Richard, 150
Germany, 39, 41
Gilmore, Ruth Wilson, 34
Google, 12, 86, 95, 111, 113, 122, 154,
  158, 200n11
Gore, Al
  bootstrapping and, 150–151
  charter schools and, 120
  Clinton and, 4, 11, 29, 36, 38–46,
    49–50, 53, 60, 84, 120, 150–151
  human capital and, 43
  libraries and, 84
  new economy and, 29, 36–46, 49–50,
    53
  pivots and, 60
Governance, 22–23, 25, 44, 112, 144,
  158, 183, 197n4, 200n12
Gramsci, A., 38, 180
Grand Central Partnership, 22

Gray, Vincent, 64, 76, 119, 163, 181
Great Society, 33–35
Green Book, 54
Greenspan, Alan, 10
Grow with Google, 158
Gruber, Frank, 65
Gusman, Patrick, 30

H-1B visas, 9
Hackathons, 65, 96
Hammond, Allen, IV, 45
Hard Hat Riot, 205n4
Harris, Donna, 64
Harris, Kamala, 4
Harvey, David, 22–23, 179
Health care
  disabilities and, 14, 32, 34, 115, 203n1
  home, 9
  new economy and, 31, 33, 195n3
  public rights and, 17
Hearth, 63, 69, 194n9, 195n1
Hegemony, 39, 124, 180–181, 184,
  204n3
Henderson, Kaya, 119, 156, 158, 163,
  201n1
Hewlett-Packard, 41
Hidden curriculum, 121, 123–128,
  131–132, 135, 138, 150
High Performance Computing Act, 39,
  41, 45
Hipsters, 95–96, 101, 106, 150, 154
Homeless
  activists and, 3, 147, 173, 197n1
  bootstrapping and, 146–147, 159–160
  charter schools and, 115, 131
  increase of, 6
  labor market and, 12, 16
  libraries and, 12, 14, 23, 25–28,
    82–91, 97, 99, 103–104,
    107–109, 159–160, 172–176,
    183–190, 198n11, 203n8
  rights and, 159
  safe public spaces for, 25

Homeless (cont.)
  shelters and, 2, 27, 83, 89–92, 102,
    104, 134, 172
  status and, 197n1
  technology and, 3–4, 26
  urban development and, 197n1
  use of term, 197n1
Home rule, 19, 21
Homework, 27, 113, 125–129, 195n5,
  199n1
How to Win Friends and Influence People
  (Carnegie), 65
Human capital
  bootstrapping and, 143, 150, 153, 157,
    159–160, 165, 170, 191, 201n1
  charter schools and, 112, 116–117,
    120, 135, 140, 175
  Clinton and, 43, 45
  concept of, 44
  DC Public Library (DCPL) system and,
    177
  libraries and, 30, 177
  new economy and, 30–32, 36, 38–39,
    43–46, 48, 54–57
  pivots and, 60, 62, 76, 78
  theory for, 44
  welfare and, 44, 55, 60, 117, 159, 170
Hurricane Katrina, 142

Ideal worker, 35, 181–186
InCrowd, 16, 27, 96
  beta context of, 73–74, 106
  bootstrapping and, 24–25, 59–61, 79,
    115, 145–146, 165, 184, 194n10
  business model of, 59, 61–62, 74–75,
    121, 129, 139, 179, 182, 184, 194n10
  cultural production of, 107
  emotional labor and, 69–72
  equity and, 65–66
  ideal workers and, 182, 184
  pivots and, 59–62, 65–79, 84, 145–
    146, 165, 195n1
  as pseudonym, 194n9, 195n1

  rent and, 59, 65
  runway and, 62
  Salesforce and, 16, 73–74, 112, 130
  social reproduction and, 179, 182, 184
  teams of, 67, 69–75
  technology and, 6, 75, 115, 130, 182
Industrial Revolution, 117, 142
Inequality
  bootstrapping and, 162, 169
  capitalism and, 5, 10, 13–14, 27–28,
    31, 43, 47–48, 54–55, 125, 162, 169,
    191, 193n2
  charter schools and, 125
  digital divide and, 31 (*see also* Digital
    divide)
  new economy and, 31, 43, 47–48,
    54–55
  poverty and, 31 (*see also* Poverty)
  taxes and, 10, 162
Inflation, 31, 37
Information Age, 31, 41, 46, 155, 198n7
Instagram, 70
Institute of Museum and Library
  Services (IMLS), 95, 108
Institutional ethnography, 26–27
Insurance, 11, 32–34, 87, 92, 187, 190
Internet
  access doctrine and, 2 (*see also* Access
    doctrine)
  average speed of connecting to, 52–53
  bootstrapping and, 161, 164, 175
  broadband, 4, 14, 29, 52–53, 133,
    195nn4,5
  charter schools and, 126, 129
  Children's Internet Protection Act
    and, 198n5
  entrepreneurs and, 25, 58, 62, 179
  High Performance Computing Act
    and, 39, 41, 45
  impact of, 1–7, 12, 15, 21, 25, 191,
    193n2
  libraries and, 53–54, 82, 87, 91–92, 98
  modems and, 31, 43, 45–46

new economy and, 29–32, 39–55, 58
NSFNET and, 39
pivots and, 61–63
skills and, 15 (*see also* Skills)
universal service and, 49–51, 54
usage data on, 193n2
"Internet, The: Your Future Depends on
    It" (poster series), 1–5, 21, 161, 181,
    191
iPhones, 27, 89, 134
iSchool, 95, 101, 148, 155, 158, 162,
    198n7
iStrategyLabs, 77

Jackson, Jesse, 42
Jackson, Scoop, 150
Japan, 39, 41, 52
JavaScript, 9
Job training, 2, 8, 11, 32–35, 66, 76,
    161, 177, 196n6
Job Training Partnership Act (JTPA), 8,
    35
Johnson, Lyndon B., 35

Katz, Lawrence F., 187
Keynesianism
    Clinton and, 36, 39, 50, 55
    Democrats and, 36
    golden age and, 193n3
    labor market and, 32
    new economy and, 31–32, 36, 39, 41,
        44, 50, 55
    taxes and, 23
King, Martin Luther, Jr., 82–83, 106, 142
Kinney, Bo, 54

Labor market
    access doctrine and, 11, 13, 16, 18,
        23, 32–33, 38, 43, 96, 144, 170, 177,
        201n3
    AFSCME and, 154–155
    automation and, 31, 33, 95
    bifurcated, 23

bootstrapping and, 144, 150, 159, 165,
    177
Bureau of Labor Statistics and, 9, 151,
    195n3
capitalism and, 17 (*see also* Capitalism)
carceral state and, 34
changing skills needed in, 10–11, 13,
    176, 179 (*see also* Skills)
charter schools and, 115, 201n3
children and, 17
Clinton and, 34, 39, 41, 43, 144,
    195n3, 203n1
Comprehensive Employment and
    Training Act (CETA) and, 8
crime rate and, 203n1
DC Public Library (DCPL) and, 81
deficit-reduction measures and, 21
disabilities and, 14, 32, 34, 115, 203n1
emotional labor and, 69–72
financial crisis of 2008 and, 21
gendered wage gaps and, 6, 8, 35, 69
gentrification and, 185 (*see also*
    Gentrification)
H-1B visas and, 9
homeless and, 12, 16
hypercompetition and, 9
ideal worker and, 35, 181–186
job training and, 2, 8, 11, 32–35, 66,
    76, 161, 177, 196n6
Keynesianism and, 32
Latinx people and, 20, 22
libraries and, 93
manufacturing and, 8, 40–41
mass firings of teachers and, 118–119
meritocracy and, 149–153, 181,
    202n6, 203n7
middle class and, 1, 20–21
migrants and, 6, 10, 20, 181
neoliberalism and, 17
new economy and, 31, 34–36, 38,
    42–44
Obama and, 9, 11
outsourcing and, 9, 31, 67

Labor market (cont.)
  payroll taxes and, 44
  pensions and, 17, 32, 96, 112, 116
  pink-collar workers and, 87
  pivots and, 62
  police and, 179
  prison and, 34, 39, 179, 183, 203n1
  productivity and, 6, 35, 44–45, 55, 60,
    104, 125, 131, 152, 163
  professionals and, 10 (see also
    Professionals)
  punishment and, 175, 178
  reform and, 5, 16, 32
  retirement and, 82, 179
  segregation of, 15–19
  social reproduction and, 16–18,
    21–23, 26
  Social Security and, 44, 87
  stratification of, 18
  strikes and, 186–188, 190
  underemployed and, 51, 173
  unemployment and, 6, 11, 14, 20, 29,
    32–34, 92, 173, 203n1
  unions and, 14, 33, 41, 96, 118–119,
    142, 145, 151, 154–155, 158, 183,
    185, 187–188, 190
  venture labor and, 62
  violence and, 178, 180
  wages and, 6–10, 16–17, 20–21, 29, 31,
    43–44, 67, 121, 174, 176, 178, 180,
    187, 195n3, 201n3, 203n1
  women and, 17–18, 21, 176, 178, 187
  workfare and, 35, 56, 179
Lafer, G., 8
Latinx
  bootstrapping and, 147, 156, 168
  charter schools and, 112–113, 117,
    119, 125, 138, 185
  labor market and, 20, 22
  libraries and, 87, 101
  new economy and, 47
Lean production, 60
Lean Startup (Ries), 60, 65

Libraries
  access doctrine and, 25, 84–85, 89–90,
    95–97, 101, 189
  AFSCME and, 154–155
  American Library Association and, 87
  Black people and, 81–83, 87, 89, 93,
    96, 101, 103, 106–107, 185
  bootstrapping and, 15–16, 24–26, 30,
    79, 85, 89–90, 95, 97–108, 115, 120,
    126, 144–155, 158–169, 184, 189,
    194n10
  budgets and, 14, 81, 95–96, 162, 164
  Canadian, 88–89, 95
  capitalism and, 180, 197n1
  Carnegie and, 147, 163–164, 198n8
  children and, 82, 186, 198n5
  Clinton and, 87
  College of Information Services and,
    94, 155
  competition and, 89, 165–170,
    198n8
  computers and, 12, 25–26, 81–109,
    146, 159–160, 171–173, 176, 185,
    189–190, 198nn6,7, 203n8
  DC Public Library (DCPL) system and,
    4, 81–83, 86, 88, 94, 143, 177, 185,
    188–189, 197nn2,3, 198n11
  demographics on, 202n5
  digital divide and, 84, 90, 188
  drugs and, 86, 99, 102
  education and, 11–12, 81–83, 94
  entrepreneurs and, 25, 97, 102–103
  as essential, 185
  financial services and, 188
  free internet and, 53
  Friends of the Library and, 82, 103,
    107, 173, 186, 189, 198n11
  funding and, 14, 34, 51, 53–54, 81–83,
    90, 94–95, 146–147, 149, 153, 161,
    164–165, 179, 198nn5,8
  gender and, 87
  gentrification and, 81, 108, 172
  Gore and, 84

hipster contingent and, 95–96, 101,
  106, 150, 154
homeless and, 12, 14, 23–28, 82–91,
  97, 99, 103–104, 107–109, 159–160,
  172–176, 183–190, 198n11, 203n8
human capital and, 30, 177
importance of, 29–30
Institute of Museum and Library
  Services (IMLS) and, 95, 108
internet and, 53–54, 82, 87, 91–92, 98
labor market and, 93
Latinx people and, 87, 101
master of library science (MLS) degree
  and, 94–96, 155, 198n7
meritocracy and, 149–153, 181,
  202n6, 203n7
middle class and, 86–87
migrants and, 87
mission ambiguity and, 160–165
MLK (*see also* Martin Luther King Jr.
  Memorial Library—Central Library)
neoliberalism and, 87, 97, 100,
  198n10
philanthropies and, 94
phones and, 86, 89, 91, 99, 101,
  103–104, 172
police and, 2, 28, 82, 88, 91–92,
  98–104, 171–173, 189, 197n2
poor condition of, 82
pornography and, 86, 92, 97–98,
  100–101, 104, 134, 159, 161, 167,
  172
poverty and, 85, 90, 186
professionals and, 5, 12, 25, 27, 61,
  85–108, 144, 148, 150–160, 164,
  166, 169, 175, 181–188, 198n9
public service mission of, 89
recession and, 14, 95
reform and, 82, 100, 106
regulation and, 85–87, 91, 94
research methodology on, 26–28,
  197n2
revenue and, 95

segregation and, 102, 109, 168
skills and, 3–6, 11–15, 18, 25, 30, 35,
  44–45, 54–55, 60–61, 83–90, 93–96,
  104–107, 115, 125, 148–152, 159,
  163, 166, 177–181, 185
sleeping in, 2, 84, 99, 104–106, 159,
  164, 167, 171–172, 176, 180, 185,
  188
software and, 92, 103, 107
startups and, 25, 84–85, 88, 93, 95–96,
  102, 109
strikes and, 186–187
taxes and, 88, 95, 178
technological professionalization and,
  153–160
transformation of, 85–90
unemployment and, 14
welfare and, 87–90, 95, 185
White people and, 14, 81–82, 86–87, 89,
  93, 96–97, 102, 107, 186–188, 198n6
women and, 18, 83, 87, 93
Lieberman, Joseph, 150
Living standards, 43, 201n3
Lobbying, 64, 82, 117, 153
Los Angeles Times, 162–163

Mandate for Change (Center for
  Education Reform), 117
Manufacturing, 8, 40–41
Mapbox, 93, 102
Martinez + Johnson, 86, 108, 197n3
Martin Luther King Jr. Memorial
  Library—Central Library (MLK)
access doctrine and, 16, 116
Advisory Neighborhood Commission
  and, 88
bootstrapping and, 16, 24–26, 30, 79,
  85, 90, 100, 104–105, 108, 115, 120,
  126, 145, 148, 167–168, 184, 189,
  194n10
competing institutional cultures and,
  167–169
Cooper and, 96

Martin Luther King Jr. (cont.)
  Creative Lab and, 108
  Digital Commons of, 12, 82–86,
    89–109, 146, 159–160, 171–173, 176,
    198n6, 203n8
  Dream Lab of, 82–83, 89, 93, 95,
    101–103, 106, 108–109, 111, 147,
    150, 161, 171, 189
  Espresso Book Machine and, 95, 107
  Fab Lab of, 83–84, 89, 102, 108–109
  free internet and, 53
  hipster contingent and, 95–96, 101,
    106, 150, 154
  ideal workers and, 182–185
  mission ambiguity and, 161–162
  police and, 2, 82, 91, 98, 102, 104, 189
  professionalizing, 90–96
  renovation of, 171–173, 186, 188–190
  research methodology on, 27, 197n2
  as space of solidarity, 190–191
  transformation of, 85–90, 106
  "Your Story Has a Home Here" and,
    189–190
Marxism, 16, 203n10, 204n3
Matchbook Learning, 142
Mather Studios, 102–103
Mecanoo Architecten, 86, 108, 197n3
Mercedes divide, 56
Meritocracy, 148–153, 181, 202n6,
  203n7
Microsoft, 1–2, 42, 93, 162
Middle class
  Black, 1, 20–21, 33, 76–77, 82, 93, 119
  bootstrapping and, 149
  charter schools and, 119
  labor market and, 1, 20–21
  libraries and, 86–87
  mobility and, 20
  pivots and, 76
  as powerful idea, 182
  social justice and, 188
  White, 19–20, 23, 76–77, 82, 86–87,
    131, 149, 188

Middle West Side Data Processing
  School, 33
Mies van der Rohe, Ludwig, 96
Migrants, 3
  bootstrapping and, 149
  creative class and, 9–10
  digital divide and, 23
  H-1B visas and, 9
  labor market and, 6, 10, 20, 181
  libraries and, 87
  new economy and, 29, 33, 37
  White, 20, 87, 181
MLK. *See* Martin Luther King Jr.
  Memorial Library—Central
  Library
Mobility
  bootstrapping and, 25, 158, 169,
    175
  charter schools and, 119
  economic, 6–7, 10–11
  education and, 15, 32, 119, 158, 162,
    169, 175
  new economy and, 32, 47
  professionals and, 21–22
  skills and, 6–7, 10–11, 15, 96, 161
  social, 15, 25, 32, 47, 96, 119, 152,
    158, 161–162, 169, 175
Modems, 31, 43, 45–46
Morrisett, Lloyd, 45
Moser, Michele, 162

National Education Summit, 42
National Information Infrastructure
  (NII), 38–41, 48–50
National Telecommunications and
  Information Administration (NTIA),
  29, 31, 45–51
Nation at Risk report, 42
Native Americans, 31
Neff, Gina, 62, 73
Neoliberalism
  Atari Democrats and, 24, 30, 37,
    40–43, 52, 145, 151, 183

bootstrapping and, 21, 24, 144, 154–155, 160, 165, 169, 173, 177, 184, 191, 201nn2,3
charter schools and, 117, 139, 186–187, 200n9
Clinton and, 24, 36, 51, 54, 56, 87, 144, 174, 203n1
education and, 117, 165, 169, 174–175, 186, 201n2, 203n1
labor market and, 17
laissez-faire of, 42
libraries and, 87, 97, 100, 198n10
new economy and, 24, 31–38, 44, 48–49, 51, 54, 56–57, 173–174
poverty and, 174–175
reform and, 117, 175, 186, 200n9, 201n2, 203n1
social reproduction and, 22, 34, 173–174, 177, 184, 186–187, 203n1
technology policy and, 31–38
urban development and, 21–22, 34, 57, 175, 177, 184, 191, 194n6
welfare and, 44, 87, 117, 169, 174, 198n10, 203n1
Neumann, Adam, 77
New Deal, 35–36
New economy
access doctrine and, 24, 30–33, 36–39, 43–49, 53, 55, 57, 173–174
activists and, 34, 38, 49
Atari Democrats and, 30, 37, 40–42, 52
Black people and, 31–36, 46–47, 53
budgets and, 35–40
capitalism and, 40, 42, 195n3
carceral state and, 31–32, 34, 37–40, 48, 57
children and, 174
Clinton and, 30–31, 34, 36–45, 48–57, 174, 195n3
competition and, 32, 40, 46–57, 174–175
computers and, 30–33, 40, 45–46, 48

crime and, 37, 39, 174
digital divide and, 29–32, 36, 38–52, 56–58, 180
education and, 31–32, 37, 39, 42–46, 54
entrepreneurs and, 31, 57–58
future of access and, 52–58
gentrification and, 57
Gore and, 29, 36–46, 49–50, 53
health care and, 31, 33
human capital and, 30–32, 36, 38–39, 43–46, 48, 54–57
inequality and, 31, 43, 47–48, 54–55
internet and, 29–32, 39–55, 58
labor market and, 31, 34–36, 38, 42–44
Latinx people and, 47
migrants and, 29, 33, 37
mobility and, 32, 47
national economic crisis and, 38–42
neoliberalism and, 24, 31–38, 44, 48–49, 51, 54, 56–57, 173–174
Obama and, 11
philanthropies and, 47
phones and, 30, 45, 48–51
pivots and, 37–38
police and, 37, 48
political economy and, 34, 37, 183
poverty and, 24, 30–39, 42–49, 53–57, 183
prison and, 34, 37–40, 48
professionals and, 30, 38, 41–42, 195n3
punishment and, 31, 34, 35–37, 44, 49, 52, 55–57
recession and, 29, 36–38, 41, 54–56
recruitment and, 41, 48
reform and, 32–33, 51, 55
regulation and, 31, 34, 38, 40–41, 48–53, 56, 195n2
revenue and, 54
skills and, 30–39, 42–47, 51–57
social reproduction and, 34–35, 57

New economy (cont.)
  software and, 40, 195n3
  startups and, 29–30, 58
  taxes and, 44, 49, 51, 54, 57
  urban development and, 194nn6,7
  violence and, 37, 39
  wages and, 29, 31, 43–44
  welfare and, 30, 32, 34–37, 42, 44, 55
  White people and, 30–31, 34, 36, 42, 46
  women and, 35
New Schools Venture Fund, 141, 147, 153, 157–158, 201n1
New Teacher Project, 119
New York Times, 33
Next Generation Internet Initiative, 49
Next Gen Schools, 143
Niles, Jenny, 141
Nixon, Richard, 34–35
NoMa Business Improvement District, 108
NSFNET, 39

Obama, Barack, 87
  access doctrine and, 11
  bootstrapping and, 142–143
  education and, 4, 9, 117–118, 142
  labor market and, 9, 11
  new economy and, 11
  private schooling of, 77
  Startup America and, 64
  STEM subjects and, 9
  TechHire and, 9
  technology and, 4, 11
Ocasio, W., 203n10
Office of the Chief Technology Officer (OCTO), 29–30, 57, 133
Open-source software, 164
Organization for Economic Cooperation and Development (OECD) countries, 52
Outsourcing, 9, 31, 67

PARCC test, 114, 127, 140, 199n6
Pell Grants, 37, 48
Pensions, 17, 32, 96, 112, 116
Personal Responsibility and Work Opportunity Act (PRWOA), 37
Pew Research Center, 53–54
Philanthropies, 195n2, 199n6, 200n10
  Bill & Melinda Gates Foundation, 95, 118, 143
  bootstrapping and, 141, 145, 156–158, 162–165, 201n1
  Broad Academy and, 157–158
  budgets and, 14, 164
  Carnegie and, 147, 163–164, 182, 198n8
  Chan Zuckerberg Initiative, 154, 158
  CityBridge Foundation, 141, 143–144, 148, 153–154, 156–157, 200n12, 201n1
  Du Bois and, 16, 26, 112, 118, 200n12
  Ford Foundation, 158
  libraries and, 94
  new economy and, 47
  NewSchools Venture Fund, 141, 147, 153, 157–158, 201n1
  Rockefeller and, 182
  venture, 158
  Walton Family Foundation, 147, 157
Phones
  Android, 134
  discipline and, 133–138
  as distractions, 14, 113–115, 120–130, 133–138
  education and, 4, 27, 113–115, 120–130, 133–138, 142, 161, 167
  impact of, 12
  iPhones, 27, 89, 134
  libraries and, 86, 89, 91, 99, 101, 103–104, 172
  new economy and, 30, 45, 48–51
  professionals and, 14, 27, 89, 120, 134–137
  startups and, 68, 70–72

Pivots
  access doctrine and, 60–61, 66, 75
  beta context of, 73–74
  Black people and, 59, 70, 76–79
  bootcamps and, 64, 66, 72, 75
  bootstrapping and, 61, 79
  capitalism and, 60, 76
  Clinton and, 60, 77
  competition and, 64, 77
  customer segment, 61
  daily, 72–75
  digital divide and, 25, 60, 63, 79, 84
  education and, 62, 69
  emotional labor and, 69–72
  entrepreneurs and, 61–68, 71, 76, 78,
    196n5
  founders and, 62–66, 75
  gentrification and, 63, 76, 78
  Gore and, 60
  human capital and, 60, 62, 76, 78
  InCrowd and, 59–62, 65–79, 84,
    145–146, 165, 195n1
  internet and, 61–63
  labor market and, 62
  middle class and, 76
  new economy and, 37–38
  phones and, 68, 70–72
  poverty and, 60–61, 75, 79, 195n2
  professionals and, 61–63, 67
  recruitment and, 59, 64–65
  revenue and, 76–77, 79, 195n1
  runway and, 62
  segregation and, 67
  skills and, 25–26, 59–61, 64, 66, 76,
    78, 96, 115, 196n3
  software and, 59–62, 64, 67–69,
    73–74, 76
  startups and, 15, 25–26, 59–79, 84,
    115, 144–145, 160, 165, 173–176,
    194nn9,10, 195n1, 196nn3,4
  taxes and, 62, 76–79
  venture labor and, 62
  violence and, 63

  welfare and, 60
  White people and, 61, 63, 70, 75–79
  women and, 67–72, 196n3
Pokémon, 101–104
Police
  Advisory Neighborhood Commission
    and, 88
  Black people and, 22, 103, 171
  bootstrapping and, 191
  charter schools and, 115
  crime and, 193n4 (see also Crime)
  labor market and, 179
  libraries and, 2, 82, 88, 91–92, 98–104,
    171–173, 189, 197n2
  new economy and, 37, 48
  violence and, 16, 28, 37, 115, 188
Political economy
  access doctrine and, 5, 14, 22, 38, 49,
    60, 144–145, 170, 174
  carceral state and, 183
  feminist Marxist, 203n10
  new economy and, 34, 37, 183
  urban development and, 23
Pornography, 86, 92, 97–101, 104, 134,
    159, 161, 167, 172
Poverty
  access doctrine and, 5, 7, 11–16, 24,
    30–32, 38, 43, 46, 49, 53, 55, 57, 60–
    61, 75, 85, 90, 120, 143–144, 160,
    166, 169, 174, 180
  Aid for Families with Dependent
    Children (AFDC) and, 35
  Black people and, 31–33, 53, 142–143,
    175, 183
  bootstrapping and, 15–16, 24, 30, 61,
    85, 142–148, 153, 160–161, 165–166,
    169, 175, 194, 201n1
  carceral state and, 31, 183, 203n1
  charter schools and, 118, 120
  children and, 16, 35, 37, 118, 201n1
  digital divide and, 11–13, 24, 30–32,
    38–39, 43, 47–48, 90, 144, 174–175
  federal line for, 8

Poverty (cont.)
food stamps and, 92, 97, 203n1
ghettos and, 32
homeless and, 3 (*see also* Homeless)
libraries and, 85, 90, 186
neoliberalism and, 174–175
new economy and, 24, 30–39, 42–49,
53–57, 183
pivots and, 60–61, 75, 79, 195n2
reform and, 32–33, 55, 85, 118, 175
shelters and, 2, 27, 83, 89–92, 102,
104, 134, 172
skills and, 5, 7, 11–13, 15, 24, 30–36,
38, 43–44, 55, 57, 60–61, 85, 90,
144, 148, 166, 168, 174–175, 180
technology and, 5, 7, 12–13, 15, 24,
29–38, 42–44, 47–49, 53–55, 57, 61,
120, 143–147, 160–161, 165–166,
169, 173–174, 180, 191, 194n5
War on Poverty and, 44
Powell, Michael, 56
Predatory inclusion, 169
Prejudice, 152
Presence bleed, 127–133
Prison
carceral state and, 34, 37–40, 48, 179,
183, 191, 203n1
Clinton and, 39–40, 203n1
crime waves and, 191
labor market and, 34, 39, 179, 183,
203n1
new economy and, 34, 37–40, 48
Pell Grants and, 37
Productivity, 6, 35, 44–45, 55, 60, 104,
125, 131, 152, 163, 193n3
Professionals
Black, 18–19, 87, 89, 93, 119, 125,
138, 143, 150, 183, 185
bootstrapping and, 143–145, 148–166,
169, 203n7
certifications and, 1–2, 7, 12, 151
charter schools and, 119–120,
124–126, 128, 131, 134–139

class composition of, 150
degrees and, 6, 9, 86, 94, 149, 151–
156, 169, 197n2, 198n7, 201n3
digital divide and, 12, 90, 174, 188
education and, 7, 27, 41–42, 62, 100,
125, 128, 139, 145, 148, 150, 153,
156, 158–159, 162, 164, 175,
182–183, 186–187
hegemony and, 204n3
helping, 5, 12, 27, 85, 87, 100, 143–144,
149–150, 158–159, 166, 182, 184–190
ideal workers and, 35, 181–186
libraries and, 5, 12, 25, 27, 61, 85–108,
144, 148, 150–155, 158–160, 164,
166, 169, 175, 181–188, 198n9
master of library science (MLS) degree
and, 94–96, 155, 198n7
mobile, 21–22
networks and, 26, 156, 158, 162, 175
new economy and, 30, 38, 41–42,
195n3
PhDs and, 152, 155
phones and, 14, 27, 89, 120, 134–137
pivots and, 61–63, 67
skills and, 7 (*see also* Skills)
Teach for America (TFA) and, 155–158,
168, 188
technological professionalization and,
153–160
wages and, 10, 21, 174
Washington, DC, and, 7, 18–21, 63,
87, 90, 94, 100, 102, 104, 119,
154–156, 181, 184, 186, 188
White, 17–22, 25, 27, 30, 42, 61, 63,
89, 97, 102, 107, 124–125, 131, 138,
149–153, 161, 169, 175, 181, 183,
188, 198n9, 203n7
Public goods, 44–45, 167
Public-private partnerships, 9, 30–31,
38, 42, 64, 116, 140, 142, 167
Punishment
bootstrapping and, 165
labor market and, 175, 178

new economy and, 31, 34–37, 44, 49,
52, 55–57
police and, 191 (*see also* Police)
welfare and, 35–37, 36, 174
Python, 9, 93

Qualified High Technology Company,
79

Race to the Top grants, 147
Racism
in education, 137
institutional, 152
international, 193n3
reconceptualism and, 199n2
technology and, 14
welfare queen and, 36
Rainbow Coalition, 42
RAND Corporation, 32
Ray, Victor, 202n4
Reagan, Ronald, 8, 35–36, 39, 41–42, 51
Recession
bootstrapping and, 164, 167, 179, 183
budgets and, 19, 37, 164
charter schools and, 118, 129
education and, 14
federal government and, 19
financial crisis of 2008 and, 6, 14,
18–21, 29, 56, 84, 95, 183
libraries and, 14, 95
new economy and, 29, 36–38, 41,
54–56
postrecession growth and, 20, 129,
164, 179
taxes and, 19, 54, 95, 179
White migration and, 20
Reconceptualizing Early Childhood
Education (RECE), 116, 130, 135,
140
Recovery Act, 53
Recruitment, 88
bootstrapping and, 156
Clinton and, 41, 48

elites and, 41
entrepreneurs and, 22, 64–65
new economy and, 41, 48
pivots and, 59, 64–65
skills and, 10, 59
Reform
bootstrapping and, 141, 146, 148, 151,
154, 156–158
charter schools and, 117–118, 120
CityBridge Foundation and, 141,
143–144, 148, 153–154, 156–157,
200n12, 201n1
Democrats for Education Reform and,
117
education and, 117–118, 120,
141, 154–158, 175, 186, 200n13,
201nn1,2, 203n1
job training and, 32–33
labor market and, 32
libraries and, 82, 100, 106
neoliberal, 117, 175, 186, 200n9,
201n2, 203n1
new economy and, 32–33, 51, 55
NewSchools Venture Fund and, 141,
147, 153, 157–158, 201n1
poverty and, 32–33, 55, 85, 118, 175
Regulation
bootstrapping and, 145, 165
charter schools and, 118, 125,
139–140
deregulation and, 31, 34, 38, 41, 48,
51–52, 56, 165
libraries and, 85–87, 91, 94
new economy and, 31, 34, 38, 40–41,
48–53, 56, 195n2
Republicans, 36, 41
Retirement, 82, 179
Revenue
bootstrapping and, 15, 145, 164–165,
169
libraries and, 95
new economy and, 54
pivots and, 76–77, 79, 195n1

Revenue (cont.)
taxes and, 10, 19, 21, 76–77, 79, 95, 145, 194n6
Washington, DC, and, 1, 19, 21, 77, 79, 95, 194n6, 200n12
Reyes-Gavilan, Richard
Advisory Neighborhood Commissions (ANCs) and, 197n4
Cooper and, 189
DC Public Library (DCPL) system and, 29–30, 86, 88, 94, 96, 150, 189, 197n4, 198n11
Friends of the Library and, 198n11
meritocratic model and, 150
mission ambiguity and, 162
Rhee, Michelle, 118–119, 142, 156, 158, 163, 199n6, 200n10, 201n1
Ries, Eric, 60–61
Rivera, Marialena D., 157
Rockefeller, John D., 182
Role modeling, 66, 121, 123
Roosevelt, Franklin D., 35, 39
Route 128, 41, 151
Rubinstein, Ross, 162
Runway, 62

Salesforce, 16, 73–74, 112, 130
Samouha, Aylon, 142–143, 154
SAT scores, 42, 114, 134, 140
Sawyer, Steve, 155
SchoolForce
academic culture and, 112–116, 121, 124, 128, 130–133, 136–138
Acumen Solutions and, 112, 199n1
discipline and, 112, 114–115, 121, 132–134, 136–137
Du Bois and, 16, 111–116, 121, 124, 128, 130–138, 185, 190–191
experimentation of, 114–116
increased surveillance by, 114
Work Hard Grades and, 136–137
Scientific American journal, 39–40

Scientific-Technical Employment Program (STEP), 33
Scott, George A., 116
Scott, Janelle, T, 157–158, 201n2
Scott, W. Richard, 145
Sculley, John, 41
Segregation
bootstrapping and, 160, 162, 168, 203n1
crime and, 203n1
labor market and, 17–19, 115
libraries and, 102, 109, 168
pivots and, 67
residential, 162
Sesame Street Workshop, 45
Shelters, 2, 27, 83, 89–92, 102, 104, 134, 172
Shelton, Jim, 142–143, 154, 164
Sidwell Friends, 77
Silicon Valley, 45, 142, 151, 202n6
Skill-biased technological change (SBTC), 7–8
Skills
bootstrapping and, 13, 15, 18, 24, 30, 61, 85, 106–107, 115, 141–144, 148–152, 159, 161, 166, 168, 170, 175, 177, 202n6
charter schools and, 115–116, 120, 131
coding, 4–5, 9, 11–12, 40, 68, 75, 155, 179, 188
conceptual issues and, 7–9, 204n2
critical thinking, 8–9
digital divide and, 1, 6–13, 15, 24–25, 30–31, 38, 43, 47, 52, 60, 90, 116, 144, 170, 174–176
gaps in, 7–11, 15, 42, 66, 116, 175–176
gender and, 8, 32
H-1B visas and, 9
ideal worker and, 35, 181–186
individual, 1, 7, 10, 180
job training and, 2, 8, 11, 32, 35, 66, 76, 161, 177, 196n6

libraries and, 3–6, 11–15, 18, 25, 30, 35, 44–45, 54–55, 60–61, 83–90, 93–96, 104–107, 115, 125, 131, 148–152, 159, 163, 166, 177–181, 185

mobility and, 6–7, 10–11, 15, 96, 161

new economy and, 30–39, 42–47, 51–57

occupational data on effects of, 176–177

pivots and, 25–26, 59–61, 64, 66, 76, 78, 96, 115, 196n3

poverty and, 5, 7, 11–13, 15, 24, 30–38, 43–44, 55, 57, 60–61, 85, 90, 144, 148, 166, 168, 174–175, 180

productivity and, 6, 35, 44–45, 55, 60, 104, 125, 131, 152, 163

recruitment and, 10, 59

shortages in, 7, 11

standardized testing and, 42–43, 111, 117, 122, 127, 134, 140, 153, 199n8, 203n1

startups and, 16, 59–61, 78, 93, 115, 143–144, 148, 166, 175, 181, 196n3

STEM, 9, 15, 83, 176

technology and, 1–9, 13–18, 24, 26, 30–31, 43–44, 47, 55, 57, 61, 66, 83–86, 89, 95, 115, 131, 141, 143, 149, 159, 166, 176–177, 180, 202n6

training and, 1, 5–18, 24, 26, 33–39, 42–43, 54, 64, 66, 78, 85–86, 89, 95, 149–151, 175–179

unemployment and, 11, 33, 78

wages and, 7–10, 43, 174, 176

Skills-to-job pipeline, 150, 152

Sloan, Van, 33

Social justice, 175, 188

Social media, 70, 89, 107, 135, 137, 141

Social reproduction

bootstrapping and, 144, 146, 166–169, 173, 177, 180

capitalism and, 16–17, 177–182, 184, 195n3

education and, 16–18, 21, 26, 34, 144, 146, 169, 179, 181, 186–187, 190, 203n1

InCrowd and, 179, 182, 184

intraclass antagonism and, 185

labor market and, 16–18, 21–23, 26

Marxist feminists and, 16

neoliberalism and, 22, 34, 173–174, 177, 184, 186–187, 203n1

new economy and, 34–35, 57

new mode for, 178–181

problem of access and, 173–174

Social Security, 44, 87

Software

bootstrapping and, 164–165, 174, 176, 182, 184, 202n6

charter schools and, 112, 122, 200n11

coding and, 4–5, 9, 11–12, 40, 68, 75, 155, 179, 188

libraries and, 92, 103, 107

new economy and, 40, 195n3

startups and, 59–62, 64, 67–69, 73–74, 76, 196n3

South Korea, 52

Soviet Union, 38–40, 49

Stagflation, 33, 35

Standardized tests, 42–43, 111, 117, 122, 127, 134, 140, 153, 199n8, 203n1

Startup America, 64

Startup Middle School, 30

Startup Mixology: Tech Cocktail's Guide to Building, Growing, & Celebrating Startup Success (Gruber), 65

Startups

2U, 142

beta context of, 73–74

bootcamps and, 64, 66, 72, 75

bootstrapping and, 15, 24–25, 58, 61, 115, 142–149, 153–154, 159–162, 165–166, 169, 175–176, 180, 184, 194nn9,10, 203n9

Startups (cont.)
  capital and, 17, 30, 62, 78, 143, 146,
    159, 175, 180–181, 184, 188
  charter schools and, 25, 115, 119,
    139
  computers and, 65, 73
  cultural labor and, 196n3
  data management platforms and, 16
  DC Tech (promotional network), 77,
    83, 161–162, 181
  DC tech (sector), 64–65, 68, 70, 72,
    75–79, 83, 161–162, 181, 184
  Du Bois and, 16, 24–25, 115, 119,
    139, 145, 148–149, 175, 184, 188,
    194nn9,10
  emotional labor and, 69–72
  entrepreneurs and, 16–17, 25, 58,
    61–62, 65, 68, 76, 78, 143, 149, 154,
    159, 176, 179
  equity and, 64–66
  gentrification and, 63, 76, 78, 181
  hackathons and, 65, 96
  InCrowd, 16 (see also InCrowd)
  libraries and, 14–15, 25, 84–85, 88, 93,
    95–96, 102, 109
  new economy and, 29–30, 58
  organizational culture and, 66–67
  phones and, 68, 70–72
  pivots and, 15, 25–26, 59–79, 84,
    115, 144–145, 160, 165, 173–176,
    194nn9,10, 195n1, 196nn3,4
  rent and, 59, 65
  research methodology on, 26–28
  runway and, 62
  skills and, 16, 59–61, 78, 93, 115,
    143–144, 148, 166, 175, 181, 196n3
  software and, 59–62, 64, 67–69, 73–74,
    76, 196n3
  venture labor and, 62
Startup Weekend EDU, 143
State of the Union, 55, 174
Status, 23, 47–48, 69, 123, 153, 184,
  197n1

STEM (science, technology, engineering,
  and math), 4, 9, 15, 77, 83, 117, 119,
  165, 176
Stevenson, Siobhan A., 88–89, 95
Stovall, Pamela, 188–189
Straubhaar, Joseph, 54
Strikes, 186–188, 190
Subsidies, 32, 45, 51, 70, 79, 89, 191
Suburbs, 2, 20–23, 31, 41, 46, 101, 118,
  151
Switzerland, 53

Task Force on the Future of the DC
  Public Library System, 82
Taxes
  bootstrapping and, 145, 151, 162,
    168
  Clinton and, 4, 151
  credits and, 78
  inequality and, 10, 162
  Keynesianism and, 23
  libraries and, 88, 95, 178
  new economy and, 44, 49, 51, 54, 57
  payroll, 44
  pivots and, 62, 76–79
  recession and, 19, 54, 95, 179
  reduction of, 76
  revenue and, 10, 19, 21, 76–77, 79, 95,
    145, 194n6
  revolts and, 14
  Washington, DC, and, 19, 21, 44,
    77–79, 88, 95, 194n6
Tax increment financing (TIF), 194n6
Teach for America (TFA), 128, 155–158,
  168, 188
Tech Cocktail, 64, 65
TechHire, 9
Technology
  academic culture and, 121–133
  access doctrine and, 18 (see also Access
    doctrine)
  Atari Democrats and, 24, 30, 37,
    40–43, 52, 145, 151, 183

broadband, 4, 14, 29, 52–53, 133,
    195nn4,5
Clinton and, 4, 30, 87
computers and, 1 (*see also*
    Computers)
DC's digital district and, 18–24
digital divide and, 176 (*see also* Digital
    divide)
as distraction, 113–115, 120–138
dot-com boom and, 62, 65
homeless and, 3–4, 26
InCrowd and, 6, 75, 115, 130, 182
internet and, 29–32, 39–55, 58
    (*see also* Internet)
modems and, 31, 43, 45–46
Obama and, 4, 11
Office of the Chief Technology Officer
    (OCTO) and, 29–30, 57, 133
phones and, 4 (*see also* Phones)
poverty and, 5, 7, 12–13, 15, 24,
    29–38, 42–44, 47–49, 53–55, 57,
    61, 120, 143–147, 160–161,
    165–166, 169, 173–174, 180, 191,
    194n5
presence bleed and, 127–133
skills and, 1–9, 13–18, 24, 26, 30–31,
    43–44, 47, 55, 57, 61, 66, 83–86, 89,
    95, 115, 131, 141, 143, 149, 159,
    166, 176–177, 180, 202n6
STEM and, 4, 9, 15, 77, 83, 117, 119,
    165, 176
Technology for America's Economic
    Growth, 40
TED Talks, 88, 180
Telecommunications Act, 50–52
Temporary Assistance for Needy
    Families (TANF), 37
Thornton, Patricia H., 203n10
TrackMaven, 77
Trujillo, Tina, 157
Trump, Donald, 195n4
Trump, Ivanka, 4
Twitter, 71, 125

Underemployed, 51, 173
Unemployment, 6, 8, 11, 14, 20, 29,
    32–34, 92, 203n1
United Teachers of Los Angeles, 187
Universal service, 49–51, 54
Urban development
    housing and, 197n1
    neoliberalism and, 21–22, 34, 57, 175,
        177, 184, 191, 194n6
    new economy and, 194nn6,7
    political economy and, 23
USAJobs, 4

Venture labor, 62
Video games, 4, 41, 70, 107, 111,
    125–126, 130–132, 152, 171
Vietnam War, 205n4
Violence
    bootstrapping and, 159, 167–168, 170
    capitalism and, 17
    charter schools and, 115–116, 128,
        178
    crime and, 183 (*see also* Crime)
    gentrification and, 181
    labor market and, 178, 180
    new economy and, 37, 39
    pivots and, 63
    police and, 16, 28, 37, 115, 188
Violent Crime and Law Enforcement
    Act, 37, 39

Wacquant, Loïc, 34, 37, 203n1
Wages
    Ford and, 180
    gaps in, 20–21
    labor market and, 6–10, 16–17, 20–21,
        29, 31, 43–44, 67, 121, 174, 176,
        178, 180, 187, 195n3, 201n3, 203n1
    new economy and, 29, 31, 43–44
    "Plan, The" and, 21
    professionals and, 10, 21, 174
    skills and, 7–10, 43, 174, 176
    Washington, DC, and, 20

Waiting for "Superman" (documentary), 118

Waivers, 35

Walton Family Foundation, 147, 157

War on Poverty, 44

Washington, DC
access doctrine and, 7, 18–26, 30, 32, 53, 57, 75, 95, 150, 184–185
"Bridging the Digital Divide in the District of Columbia" and, 29
budget of, 19, 81
Chamber of Commerce and, 29, 78, 180
crime and, 76, 191
DC Public Charter School Board (DCPCSB) and, 112, 116–117, 147
DC Public Library (DCPL) system and, 4, 81–83, 86, 88, 94, 143, 177, 185, 188–189, 197nn2,3, 198n11
DC Public Schools (DCPS) system and, 116, 134, 139, 200n12, 201n1
Department of Human Services (DHS) and, 172
Digital DC and, 76–77, 181
eight wards of, 19
gentrification and, 18, 63, 76, 78, 81, 108, 172, 181, 197n4
Gray and, 64, 76, 119, 163, 181
home rule and, 19, 21
job growth in, 20
job training reputation of, 196n6
Office of the Chief Technology Officer (OCTO) and, 29–30, 57, 133
"Plan, The," and, 21
professionals and, 7, 18–21, 63, 87, 90, 94, 100, 102, 104, 119, 154–156, 181, 184, 186, 188
revenue and, 1, 19, 21, 77, 79, 95, 194n6, 200n12
suburbs and, 2, 20–23, 31, 41, 46, 101, 118, 151
taxes and, 19, 21, 44, 77–79, 88, 95, 194n6

teacher demographics and, 202n5
wages and, 20
Williams and, 20, 81–82, 118
Washington Post, 29, 81, 150, 152, 185
Washington Teachers Union, 118
W. E. B. Du Bois Public Charter High School (Du Bois)
academic culture and, 112–116, 121–133, 136–138
access doctrine and, 116
AP for All and, 163
beta operations of, 106
bootstrapping and, 24–26, 112, 115, 120, 126, 137, 139, 145–146, 148, 158, 167–168, 175, 184, 194n10
Carroll and, 111, 113–114, 122, 125–126, 130, 136, 168, 193n1
competing institutional cultures and, 167–169
DC-CAS test and, 114, 140
DC Public Charter School Board (DCPCSB) and, 112, 116–117, 147
discipline and, 16, 112, 114–116, 121, 132–138
experiments of, 111, 114–121, 132, 138–140, 201n14
hidden curriculum and, 121, 123–128, 131–132, 135, 138, 150
homework and, 113, 125–129
ideal workers and, 182–184
lost computers of, 122
mission ambiguity and, 161, 163
philanthropies and, 16, 26, 200n12
presence bleed and, 127–133
as pseudonym, 25
Reconceptualizing Early Childhood Education (RECE) and, 116, 130, 135, 140
reform and, 120
research methodology on, 27–28
SchoolForce and, 16, 111–116, 121, 124, 128, 130–138, 185, 190–191
as space of solidarity, 190–191

---

startups and, 16, 24–25, 115, 119, 139, 145, 148–149, 175, 184, 188, 194n10
STEM subjects and, 117, 119
teaching team of, 111–112
TFA and, 156, 158
Think Tank of, 111, 113, 123, 125–127, 130, 135, 137–138
as turnaround school, 138–140
Video Game Design and, 130–132
Work Hard Grades and, 136–137
Welfare
    Black people and, 32, 35–36, 185, 193n4
    bootstrapping and, 144, 159, 165, 167, 169–170, 176, 201n3, 203n1
    carceral state and, 60, 169–170, 198n10, 203n1
    charter schools and, 117
    Clinton and, 34, 37, 42, 55, 60, 87, 174, 203n1
    ending, 34–35
    food stamps and, 92, 97, 203n1
    human capital and, 44, 55, 60, 117, 159, 170
    JTPA and, 8
    Keynesian, 23
    libraries and, 87–90, 95, 185
    neoliberalism and, 44, 87, 117, 169, 174, 198n10, 203n1
    new economy and, 30, 32, 34–37, 42, 44, 55
    pivots and, 60
    postwar economic boom and, 193n4
    punishment and, 35–37, 174
    waivers and, 35
    War on Poverty and, 44
    workfare and, 35, 56, 179
Welfare queens, 36
WeWork, 64, 77–78, 196n5
"What's Wrong with American High Schools" (Gates), 162–163
White flight, 19, 23
White House, 4, 36, 64, 118, 152

White people
    bootstrapping and, 146, 149–153, 156–157, 161, 169
    charter schools and, 112–113, 118, 124–125, 138, 140
    digital district of, 18–24
    entrepreneurs and, 22–23, 25, 64, 149–150, 157, 179
    gentrification and, 22, 63, 76, 181
    labor market and, 19–22
    libraries and, 14, 81–82, 86–87, 89, 93, 96–97, 102, 107, 186–188, 198n6
    middle class, 19–20, 23, 76–77, 82, 86–87, 131, 149, 188
    migrants and, 6, 20, 87, 181
    new economy and, 30–31, 34, 36, 42, 46
    pivots and, 61, 63, 70, 75–79
    professionals and, 17, 19, 21–22, 25, 27, 30, 42, 61, 63, 89, 97, 102, 107, 124–125, 131, 138, 149–153, 161, 169, 175, 181, 183, 188, 198n9, 203n7
    wage increases and, 20
    women, 17, 187
Wiggins, Andrea, 155
Williams, Anthony, 20, 81–82, 118
Willis-Graham Act, 50
Wilson, Darren, 127
Wingo, Henry, 29–30
Women
    Aid for Families with Dependent Children (AFDC) and, 35
    Black, 17, 21, 142–143, 147, 153
    bootstrapping and, 142–143, 147, 153, 202n5
    charter schools and, 18
    computers and, 196n3
    cultural labor and, 196n3
    emotional labor and, 69–72
    feminism and, 16, 199n2, 203n10
    gender-based wages and, 8

labor market and, 17, 21, 176, 178,
   187
libraries and, 18, 83, 87, 93
new economy and, 35
pivots and, 67–72, 196n3
White, 17, 187
Workfare, 35, 56, 179
Workforce Investment Act (WIA), 8
Work Hard Grades, 136–137
World Treaty Organization (WTO), 49

Young, John, 41
Youth Opportunity Campaign, 33
YouTube, 103, 107, 109, 125, 173

Zipper, David, 64
Zuckerberg, Mark, 196n3, 202n6